ENVIRONMENTAL SCIENCE, ENGINEERING AND TECHNOLOGY

LIQUEFACTION
ANALYSIS AND ASSESSMENT

Environmental Science, Engineering and Technology

Additional books and e-books in this series can be found on Nova's website under the Series tab.

ENVIRONMENTAL SCIENCE, ENGINEERING AND TECHNOLOGY

LIQUEFACTION
ANALYSIS AND ASSESSMENT

COREY HANSON
EDITOR

nova
science publishers
New York

Library of Congress Cataloging-in-Publication Data

ISBN: 978-1-53614-773-5

Published by Nova Science Publishers, Inc. † New York

CONTENTS

Preface		**vii**
Chapter 1	Large-Scale Hydrogen Liquefaction Processes *Majid Aasadnia, Mehdi Mehrpooya and Tatiana Morosuk*	**1**
Chapter 2	Liquefaction Hazard and Nuclear Power Plant Safety *Tamás János Katona*	**45**
Chapter 3	Simulation and Optimization of the Propane Pre-Cooled Mixed Refrigerant (C3MR) Process Using Exergy-based Method *Eko Primabudi, Liliana Calderon and Tatiana Morosuk*	**79**
Bibliography		**103**
Related Nova Publications		**201**
Index		**205**

PREFACE

Due to its high energy density, liquid hydrogen is one of the best alternative fuels. As such, *Liquefaction: Analysis and Assessment* introduces hydrogen liquefaction techniques and provides an outlook concerning common methodology. Basic and conventional cycles used for hydrogen liquefaction are reviewed, and industrial in-service hydrogen liquefaction plants are discussed.

The severe accident at the Fukushima Dai-ichi nuclear power plant demonstrated that the phenomena caused by an earthquake may actually cause more disastrous consequences than the vibratory ground motion itself, one of these phenomena being soil liquefaction. The authors propose that an analysis of the liquefaction hazards and consequences should be performed for the safety analysis of nuclear plants.

Lastly, the authors address how uncertainties in the liquefied natural gas market are growing as new producers, including the United States and Australia, emerge. In an effort to tackle these market uncertainties, process optimization is an essential part of the liquefied natural gas project. In this instance, an exergy-based method is demonstrated and assessed for both thermodynamic and cost objectives.

Chapter 1 - Liquid hydrogen (LH_2) is the best alternative fuel because of its energy density. The transportation of gaseous hydrogen requires high-pressure bulky storage tanks. Such tanks are not technically feasible because of the mechanical resistance limitations. However, LH_2 storage tanks are important for long-distances transportation as well as for large capacity LH_2 storage systems. Many special high-tech technologies, such as spacecraft in the aerospace, experiments with bubble cambers in the nuclear physics, etc. use liquid hydrogen as the strategic working fluid. Therefore, the demand for LH_2 will increase in the near future. As a result, the development of large-scale hydrogen liquefaction (LHL) plants will become increasingly important. This chapter introduces hydrogen liquefaction techniques and provides an outlook concerning common methods of liquefying hydrogen: Basic and conventional cycles used for hydrogen liquefaction are reviewed, and industrial in-service LHL plants are discussed. Innovative conceptual plants for hydrogen liquefaction are also introduced. Furthermore, hybrid liquefiers with the implementation of renewable energy sources in different configurations are presented. Finally, the exergy concept is used in order to evaluate the thermodynamic efficiency of the LHL plants.

Chapter 2 - The severe accident at the Fukushima Dai-ichi nuclear power plant demonstrated that the secondary phenomena caused by an earthquake can cause more disastrous consequences that the vibratory ground motion itself. Moreover, even if the probability of exceedance of the external hazard effects is very low compared to the plant design basis, the nuclear power plant's (NPP) safety must be ensured by appropriate severe accident measures. Soil liquefaction can be a possible secondary phenomenon of earthquakes at soil sites of NPPs. Analysis of the liquefaction hazard and liquefaction consequences should be performed for the design basis conditions, for safety analysis of the plant and for the severe accidents. The objectives and basic requirements of these analyses are given in the chapter. The soil liquefaction hazard is a matter of acceptance of the site, therefore the hazard should be excluded from the design basis. If the potential for soil liquefaction is found to be unacceptable, the site shall be deemed unsuitable unless practicable engineering solutions are demonstrated to be available for soil stabilization. The main task is to demonstrate that plant safety can be made either via integration of the earthquake-induced liquefaction consequences into seismic probabilistic safety analysis or via adaptation of this method to the case of liquefaction.

The difficulties for adaptation of the seismic probabilistic safety analysis methodology is thoroughly considered, taking into account the advances in the research of the liquefaction phenomenon and liquefaction induced building movements as well as the specific features of the structures, systems and components of the NPPs. The applicability of the state-of-the-art liquefaction risk analysis methodologies for nuclear power plants is also discussed. The possibility for assessing the nuclear power plant safety via code-deterministic margin calculation is demonstrated and could be an easy method to adapt compared to those in seismic probabilistic safety analysis. The severe accident analysis for the liquefaction case is briefly described as a logical continuation and amendment of the deterministic margin type evaluation of the safety with respect to liquefaction.

Chapter 3 - The uncertainties in the LNG market is currently growing, as the new producers, such as the United States and Australia emerges and the traditional importers such as Japan is shifting towards other fuels and renewables. Nonetheless, the growth of LNG trade is still proven to be healthy at a 0.5% increase over the last four years. Most of the LNG is traded with traditional long-term contract, as current importers, mostly in Asia, seeking for the security of supply. On the liquefaction side, Propane pre-cooled mixed refrigerant, known as C3MR, is the most dominant technology utilized in LNG plants. It is reported that 37% of the existing commercial plants used C3MR, whereas C3MR/SplitMR and AP-X® process, the extensions of C3MR design, accounts for 21% and 13%, respectively. In order to tackle the market uncertainties, process optimization is an essential part of the LNG project, giving the plant a competitive advantage in the long-term trade. In this case, exergy-based method is demonstrated on design assessment and optimization for both thermodynamic and cost objectives. The result is a high-resolution analysis of the thermodynamic performance of the system, in which the irreversibilities are revealed. Subsequently, the exergoeconomic analysis of C3MR process is implemented in depth by combining the economic analysis and exergy analysis. Finally, the optimization is achieved by combining the analyses with genetic algorithm technique in order to obtain the optimum cost of the liquefied natural gas. The design variables chosen for optimization are based on the exergy destruction data provided from the exergy analysis. It is shown that the optimization approach resulted in lower energy consumption by nearly 8%, exergetic efficiency improvement of 2.8% and a reduction of cost of product by 9.3%.

In: Liquefaction: Analysis and Assessment ISBN: 978-1-53614-773-5
Editor: Corey Hanson

Chapter 1

LARGE-SCALE HYDROGEN LIQUEFACTION PROCESSES

Majid Aasadnia[1], Mehdi Mehrpooya[1,2,*] and Tatiana Morosuk[3]
[1]Renewable Energies and Environment Department, University of Tehran, Iran
[2]Hydrogen and Fuel Cell Laboratory, University of Tehran, Iran
[3]Institute for Energy Engineering, Technische Universität Berlin, Germany

ABSTRACT

Liquid hydrogen (LH_2) is the best alternative fuel because of its energy density. The transportation of gaseous hydrogen requires high-pressure bulky storage tanks. Such tanks are not technically feasible because of the mechanical resistance limitations. However, LH_2 storage tanks are important for long-distances transportation as well as for large capacity LH_2 storage systems. Many special high-tech technologies, such as spacecraft in the aerospace, experiments with bubble cambers in the nuclear physics, etc. use liquid hydrogen as the strategic working fluid.

* Corresponding Author Email: mehrpoya@ut.ac.ir.

Therefore, the demand for LH_2 will increase in the near future. As a result, the development of large-scale hydrogen liquefaction (LHL) plants will become increasingly important. This chapter introduces hydrogen liquefaction techniques and provides an outlook concerning common methods of liquefying hydrogen: Basic and conventional cycles used for hydrogen liquefaction are reviewed, and industrial in-service LHL plants are discussed. Innovative conceptual plants for hydrogen liquefaction are also introduced. Furthermore, hybrid liquefiers with the implementation of renewable energy sources in different configurations are presented. Finally, the exergy concept is used in order to evaluate the thermodynamic efficiency of the LHL plants.

Keywords: hydrogen liquefaction, cryogenic process, exergy analysis

INTRODUCTION

Liquid hydrogen (LH_2) is a material that has the exclusive features of alternative high-energy density fuel, and has a relative low boiling point compared to other cryogenic products (Trevisani et al., 2007; Yang and Ogden, 2007). The storage of material with a high-energy content to meet the requirements of long-distance transportation are provided through the liquefaction of hydrogen (Stroman et al., 2014; Sharma and Ghoshal, 2015). Moreover, LH_2 is considered as a promising energy carrier for energy storage systems that is supposed to cover the energy peak demand within renewable energy-based electrical grids (Li, Chen and Ding, 2010; Uyar and Beşikci, 2017; Won et al., 2017; Wang et al., 2017; Al-Hallaj et al., 2017; Hemmati and Saboori, 2016; Eriksson and Gray, 2017). There are many other well-known applications for LH_2 (Denton and Nicholls, 1964). Low boiling point and high specific trust of the LH_2 makes it suitable for rockets and space vehicles (Edeskuty, 1964). LH_2 is also used in bubble chambers (Hernandez and Birmingham, 1964) and high-energy nuclear physics (Webb, 1964). LH_2 associated with clean energy systems and high-technology applications (Berstad et al., 2009; Berstad et al., 2013; Cardella et al., 2017; Cardella et al., 2017a).

The liquefaction of hydrogen is a very energy-intensive process (Wolf, 2002). Specific energy consumption (SEC) of in-service Praxair plants in the US is reported in the limit of 12.5 and 15 kWh/kg_{LH2} (Drnevich, 2003). The SEC of the large-scale plant of Linde AG (Ingolstadt, Germany) is mentioned in several publications: 13.58 kWh/kg_{LH2} (Bracha et al., 1994; Gross et al., 1994), 13 kWh/kg_{LH2} (Weindorf et al., 2003) and 15 kWh/kg_{LH2} (Quack, 2002). It should be noted that the SEC values of liquefiers by Linde AG is used as a reference value. The information about the SEC of other in-service plants around the world, e.g., Air Products & Chemicals, Air Liquide S.A., and a new Linde AG large-scale plant system in Leuna (Germany), is not available in open information sources. Based on the conducted review, it can be deduced that the SEC of commercial large-scale hydrogen liquefiers is in the range of 10 to 15 kWh/kg_{LH2}. Ohlig and Decker (2014) predicted that the SEC of future plants will be in the range of 7.5 to 9 kWh/kg_{LH2}.

During the last decade, several research groups focused their activities on the improvement of the performance of hydrogen liquefaction processes to decrease the value of SEC and/or increase the exergy efficiency (if an exergy analysis was applied). The different process configurations discussed in this chapter show the patterns and trends of the hydrogen liquefaction.

THE CHALLENGE OF THE ORTHO-PARA CONVERSION THROUGH HYDROGEN LIQUEFACTION

Hydrogen has two isomeric forms (Figure 1). The nuclear spins vectors of the two protons are coincident in ortho-hydrogen. The direction of them is different in the para-hydrogen, therefore the physical properties of the two isomers are different (Giauque and Johnston, 1928; Brickwedde; Scott and Taylor, 1935).

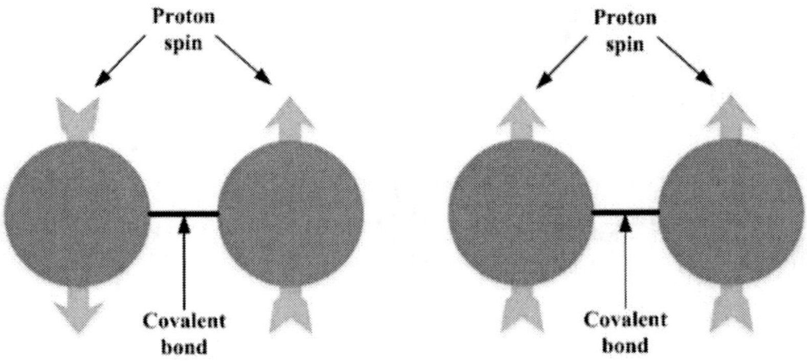

Figure 1. Spin isomers of molecular hydrogen: a) para-hydrogen anti-parallel nuclear spins with lower energy state, and b) ortho-hydrogen: parallel nuclear spins with higher energy state.

Ortho-hydrogen has a higher level of energy and naturally tends to convert to para-hydrogen (Schmauch and Singleton, 1964). Therefore, the ortho- to para-conversion is exothermic. The other thermal aspect is shown in Figure (Jensen et al., 1980). As it can be seen, the lower the temperature, the higher the equilibrium concentration of para-hydrogen. However, the equilibrium concentrations are not affected by pressure (McCarty et al., 1981).

The potential problems of LH_2 storage and transportation are examined, for example by Gursu et al., (1993); Sherif et al., (1997); Zhang et al., (2005); Notardonato (2006). The main result of research is the following: ortho-hydrogen naturally tends to gradually convert to para-hydrogen through an exothermic conversion. The conversion heat is $0.388\ \text{kcal/mol}$ while the heat of hydrogen evaporation is $0.213\ \text{kcal/mol}$. As a result, LH_2 will gradually evaporate if the storage tank has ortho-hydrogen. The so-called “boil off” phenomenon is the main challenge of LH_2 storage and handling (Tzimas et al., 2003). Therefore, ortho-hydrogen has to be converted into para-hydrogen during the hydrogen liquefaction in order to reduce negative boil-off effects, especially for long-time storage (Kalanidhi, 1988) and ortho- to para-conversion mechanisms should be considered in real hydrogen liquefaction processes (Fradkov and Troitskii, 1965; Bliesner, 2013).

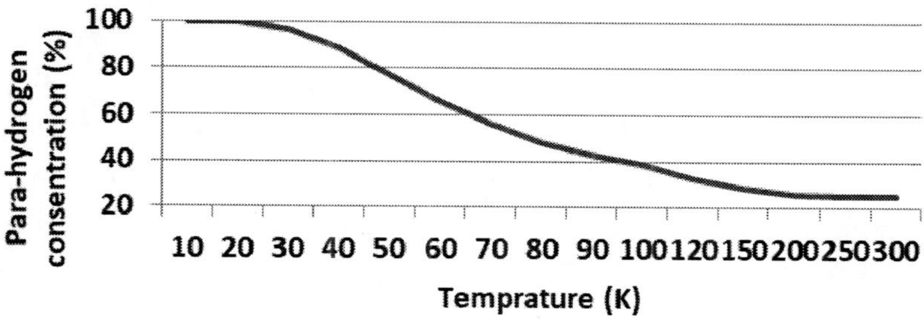

Figure 2. Equilibrium para-hydrogen concentration versus temperature.

Basic Cycles of Hydrogen Liquefaction

In the year 1898, James Dewar (UK) succeeded to liquefy hydrogen. The flow rate of the production was 4 cc/min (Dewar, 1898). Figure 3 shows the first hydrogen liquefaction "apparatus." Gaseous hydrogen was pressurized (to 180 bar), then carbonic acid and liquid air were utilized to precool it to−250°C. Dewar's flask, a vacuum insulated vessel, was the final step of the process, where hydrogen was supposed to be expanded to the atmospheric pressure. About 1% of the feed gas hydrogen could be liquefied.

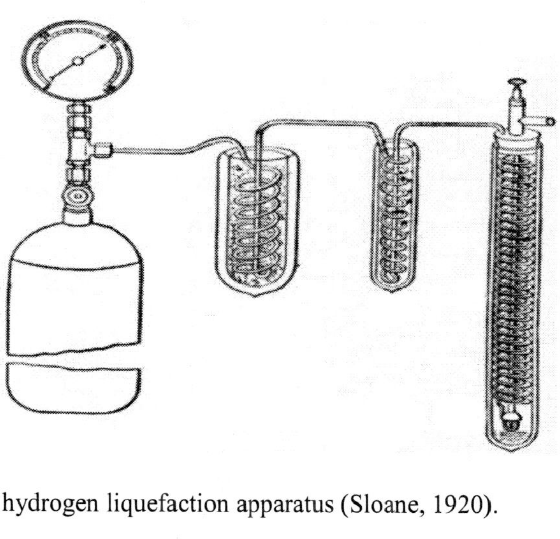

Figure 3. Dewar hydrogen liquefaction apparatus (Sloane, 1920).

The cryogenic fundamentals are discussed in detail, for example by Haselden (1971), Barron (1972), Walker (1983), Windmeier and Barron (2000), Mukhopadhyay (2010), and Walker (2012). At the first step of a gas liquefying system, the gas pressure is increased and its temperature is decreased in order to become a non-ideal gas. After that, the main process of the LHL systems is throttling. If the maximum inversion temperature of the gas is lesser than the ambient temperature, precooling will be necessary. The basic cycles that may be used for large-scale liquefaction of hydrogen are the following:

- Pre-cooled Linde–Hampson,
- Pre-cooled dual- pressure Linde–Hampson,
- Simple Claude,
- Kapitza,
- Dual pressure Claude,
- Pre-cooled simple Claude,
- Pre-cooled dual pressure Claude,
- Helium-precooled Claude, and
- Pre-cooled mixed refrigerant cycles (Barron, 1972).

Simple Cycles for Hydrogen Liquefaction

By 1895, the gas liquefaction cycles consisted of three main processes: compression, cooling, and expansion. The liquefaction was done without pre-cooling.

The liquefaction systems were developed by adding recuperative cooling processes to simple conventional three-stage processes (Davies, 1989). The liquefying gas itself is used as the coolant in the recuperative processes. The simple schematic operates according to the Claude cycle (Figure 4 is the first one used to liquefy gases (Walker, 1983).

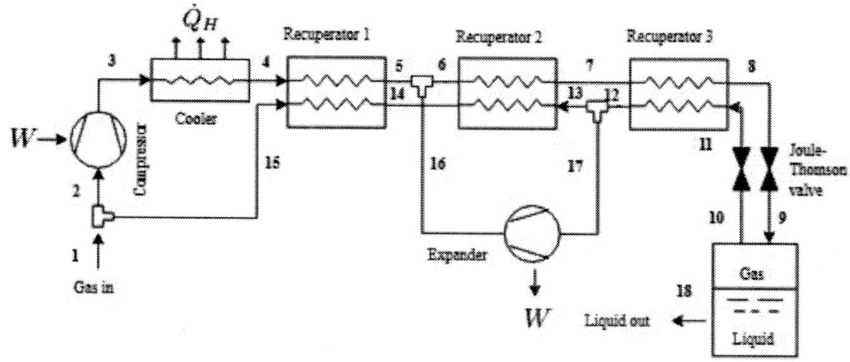

Figure 4. Conceptual design of the plant based on a simple Claude cycle (from Choudhury, 2009).

Kapitza cycle (Kapitza, 1934) is the modification of the simple Claude cycles. The third heat exchanger or low-temperature heat exchanger was eliminated and several notable practical modifications were introduced. A rotary expander was proposed instead of reciprocating one.

Dual pressure Claude cycle and Heylandt cycle are the modification of the Claude cycle (Flynn, 2005).

Precooling in Hydrogen Liquefaction Plants

The role of precooling in the cryogenic context was investigated, for example by Barron (1972), Walker (1983), Drnevich (2003), Mukhopadhyay (2010), Maytal and Pfotenhauer (2012). The liquefying gas should be precooled where its inversion temperature is lower than the ambient. Moreover, precooling results in a reduction of the enthalpy of the hot stream (h_2) for the high-temperature heat exchanger (HX-1), Figure 5 In addition, the purification can be positively affected by precooling as well (Barron, 1972).

Basic Cycles Assisted by Liquid-Nitrogen Precooling

LN_2 may be used to precool liquefying gaseous hydrogen. If the Linde-Hampson cycle is combined with a LN_2 bath, it may be used for hydrogen liquefaction (Barron, 1972). Figure shows the process flow diagram of the combined cycle. However, no pre-cooling is necessary for the Claude cycle and it can be directly utilized to liquefy hydrogen (Peschka, 2012). Precooling improves the performance of the simple Claude cycle. A combination of the dual pressure effect and the precooling feature is a modification of the simple precooled cycles (Barron, 1972).

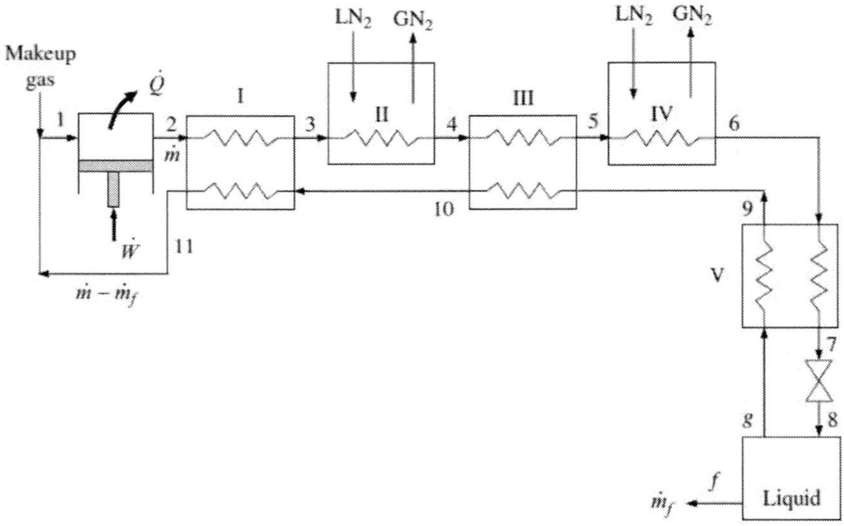

Figure 5. Pre-cooled Linde-Hampson PFD for hydrogen liquefaction (from Kanoglu et al., 2007).

Helium Pre-cooled Systems

Pre-cooling may be provided by auxiliary cycles. An option are the helium refrigeration systems that are used to precool hydrogen within the hydrogen liquefaction processes (Nandi and Sarangi, 1993). Helium itself is usually precooled by passing a liquid nitrogen bath. In fact, it is not liquefied through the process, however, precools hydrogen.

Joule-Brayton Pre-cooled Systems

Figure 6 depicts a mature modification of Figure: three auxiliary refrigeration systems assist the main Claude cycle. The auxiliary cycles that are sometimes called reverse Joule-Bryton cycles may use pure or mixed refrigerants. However, the refrigerant is not condensed through the process and just acts as the coolant.

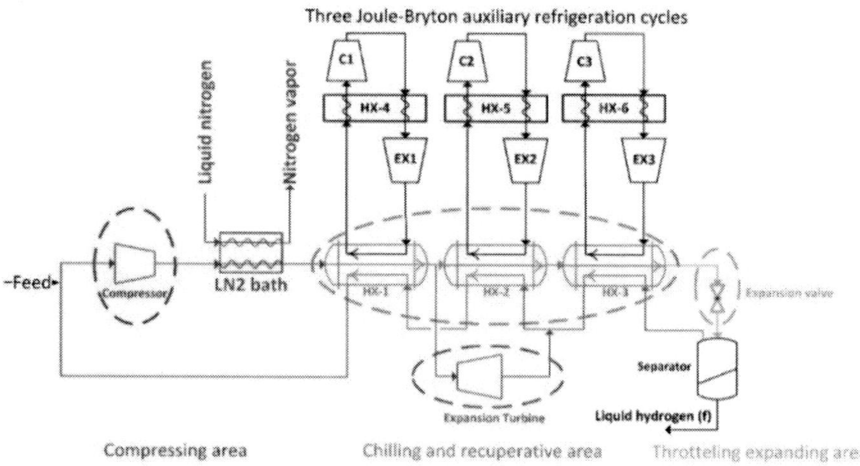

Figure 6. Conceptual design of the liquid-nitrogen and Joule-Bryton pre-cooled Claude cycle (from Asadnia and Mehrpooya, 2017).

Mixed Refrigerant Pre-cooled Systems

Mixed refrigerants can be used in a wide range of temperatures between dew to bubble temperatures while the pressure is constant (Nogal et al., 2008; Walnum et al., 2012).

In the year 1959, Kleemenko proposed to use a mixed refrigerant for a natural gas liquefaction process (Kleemenko, 1959). Figure 7 shows the Kleemenko cycle which is also the modification of the simple Claude cycle (Venkatarathnam and Timmerhaus, 2008). A phase separator and a throttling valve are used instead of the stream divider and the expander shown in Figure. The refrigerator may be easily converted to a liquefier substituting the evaporator by a liquid separator.

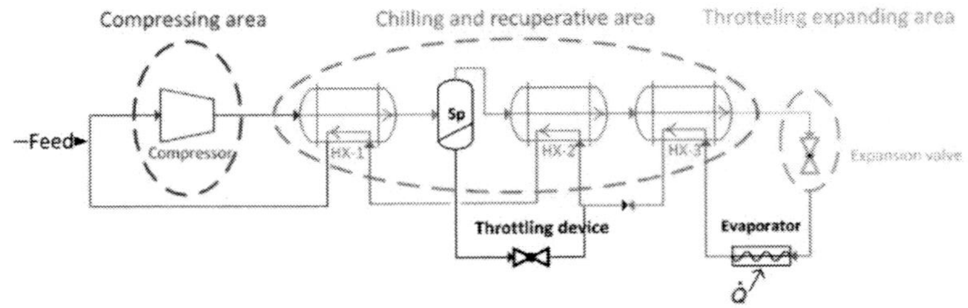

Figure 7. Kleemenko refrigerator (from Asadnia and Mehrpooya, 2017).

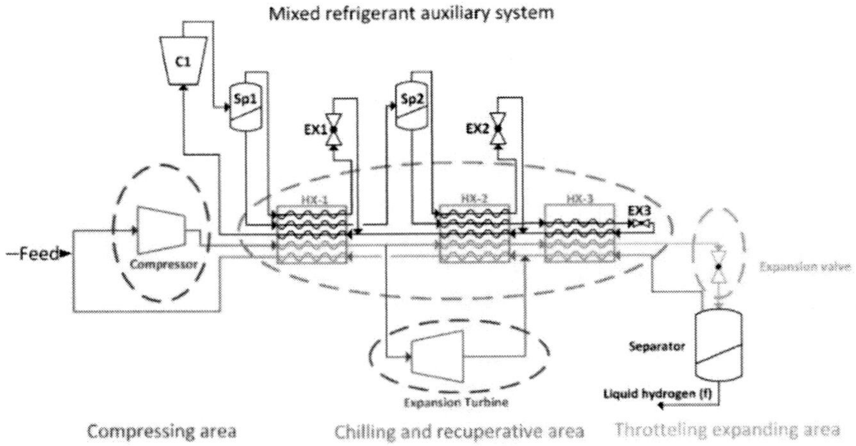

Figure 8. General configuration of the mixed refrigerant pre-cooling simple Claude cycle (from Asadnia and Mehrpooya, 2017).

Figure 8 demonstrates the common configuration of a mixed refrigerant precooled system that is a combination of the MR refrigerator and the simple Claude cycle (Flynn, 2005). The MR cycle assists the main liquefaction system. The process consists of successive phase separators that repeatedly divide the mixed refrigerant into liquid and vapor phases. The MR is cooled and partially condensed and is gradually getting heat from the forwarding streams and is heated by using the return streams.

Cascade Systems

Cascading is a well-known concept that is used as a modification of the precooling systems. Auxiliary cascade refrigeration cycles assist the main system (Barron, 1972). Thermodynamically, cascading leads to an ideal liquefaction process. Using refrigerant baths that contain liquid cryogens or use the recuperative heat exchangers is another form of cascading in gas liquefying systems that provides the most efficient and reversible process (Kerry, 2006). Furthermore, cascading or multistage cooling is a practical solution when required temperatures are in a wide range (Nogal et al., 2008). Cascading leads to an increase in the liquid yield (Walker, 1983). Table 1 lists and compares basic hydrogen liquefaction cycles. In addition, the main performance indicators of them are described briefly.

INDUSTRIAL HYDROGEN LIQUEFIERS

Table 2 lists industrial plants of hydrogen liquefaction around the world. Some of them are in service and some others have already finished their mission (West 2003; Krasae-in et al., 2010; Peschka 2012; Arend and Chelton, 2013). Their operation characteristics are usually classified and published reports may be reviewed in reference (Majid Asadnia and Mehrpooya 2017). However, Table 3 shows the reported specific energy consumption.

THE PROPOSED CONCEPTUAL PROCESSES FOR HYDROGEN LIQUEFYING

Hydrogen liquefaction processes are very energy-intensive. Details of the numerous related studies can be found, for example, in Krasae-in et al., (2010) and Asadnia and Mehrpooya (2017). Table 4 summarizes the significant features of main hydrogen liquefying processes. Table 5 shows the technical parameters and thermodynamic conditions of a typical system of hydrogen liquefying.

Table 1. Technical data of the basic hydrogen liquefaction cycles

Hydrogen liquefaction cycle	y* (%)	SEC (kWh/kg_{LH_2})	FOM (%)	Remarks
Wroblewski system (Travers, 1899)				Compressing feed hydrogen to 100 atm and pre-cooling to −183°C and expanding
Olszewski system (Rafalska‑ Łasocha, 2015)				Similar to Wroblewski, but on a large scale
Dewar system (Dewar, 1898)				Compressing feed hydrogen to 100 atm, and pre-cooling to −183°C. Capacity is 0.24 l/ht
The thermodynamically ideal liquefaction (Carnot cycle) (Barron, 1972; Peschka, 2012; Arend and Chelton, 1964)				
Feed: 21 bars, 25°C, n- GH_2				
Output: 1 bar. -253°C., $n-LH_2$	-	3.27	100	SEC = 2.18 (Krasae-in at al., 2010)
Output: 1 bar, -253°C, 99.8% $p-LH_2$	-	3.92	100	SEC = 2.89 (Krasae-in et al., 2010)
Pre-cooled ideal LH (Peschka, 2012) Output: 1 bar. -253°C., $n-LH_2$	30	16.27	20	Figure SEC = 16.24 (Vander Arend and Chelton 1964)**
Pre-cooled ideal dual pressure LH (Peschka 2012) Output: 1 bar. -253°C., $n-LH_2$	41	12.14	27	SEC = 12.12 (Vander Arend and Chelton 1964)**

Hydrogen liquefaction cycle	**y* (%)**	**SEC (kWh/kg_{LH_2})**	**FOM (%)**	**Remarks**
Pre-cooled Linde-Hampson (Nandi and Sarangi 1993) Output: 1 bar. -253°C., $n-LH_2$	12-17	72.8-79.8	4.5-5.0	Figure
Simple Claude (Asadnia and Mehrpooya 2017) Output: 1 bar. -253°C., $n-LH_2$	8	22.1	18.1	Figure Capacity = 4.46 (l/hr)
Pre-cooled ideal single pressure Claude (Peschka 2012) Output: 1 bar. -253°C., $n-LH_2$	52	10	32.6	-
Pre-cooled ideal dual pressure Claude (Peschka 2012) Output: 1 bar. -253°C., $n-LH_2$	59	8.68	37.7	SEC = 6.66 (Vander Arend and Chelton 1964)**
Pre-cooled Claude (Nandi and Sarangi 1993) Output: 1 bar. -253°C., $n-LH_2$	16-20	28-39.2	9.2-13	-
Pre-cooled dual pressure Claude (Vander Arend and Chelton 1964) Output: 1 bar, -253°C, 99.8% $p-LH_2$	-	12.26	-	-
Helium-refrigerated system (Nandi and Sarangi 1993) Output: 1 bar. -253°C., $n-LH_2$	100 ($n-H_2$) 54 ($p-H_2$)	33.6-56	6.5-11	Figure

* Liquid yield.

** Output: 1 bar, -253°C, 99.8% $p-LH_2$.

Table 2. Hydrogen liquefaction plants of the world
(West 2003; Krasae-in et al., 2010; Peschka 2012; Arend and Chelton, 2013)

Country	Location	Owner	Size (TPD*)	Year established	In operation
USA	Colorado	NBS	0.5	1952	-
USA	Ohio	APCI**	1	1956	-
USA	Painesville	Air Products	3	1957	No
USA	West Palm Beach	Air Products	3.2	1957	No
USA	Florida	APCI	3.5	1957	-
USA	California	Stearns-Roger Mfg. Co.	1.5	1957	-
USA	Florida	APCI	30	1958	-
USA	West Palm Beach	Air Products	27	1959	No
USA	Mississippi	Air Products	32.7	1960	No
USA	California	Stearns-Roger Mfg. Co.	7	1960	-
Canada	Ontario	Praxair	20	1962	Yes
USA	California	Stearns-Roger Mfg. Co.	26	1962	-
USA	California	APCI	32.5	1963	-
USA	Sacramento	Union Carbide	54	1964	No
USA	New Orleans, LA	APCI	34	1977	Yes
USA	New Orleans, LA	APCI	34	1978	Yes
Japan	Amagasaki	Iwatani	1.2	1978	No
USA	Niagara Falls, NY	Praxair	18	1981	Yes
Canada	Sarnia Ontario,	APCI	30	1982	Yes
Japan	Tashiro	MHI	0.6	1984	No
Japan	Akita Prefecture	Tashiro	0.7	1985	Yes
USA	Sacramento, CA	APCI	6	1986	Yes

Country	Location	Owner	Size (TPD*)	Year established	In operation
Japan	Tane-Ga-Shima	Japan Liquid Hydrogen	1.4	1986	Yes
Japan	Oita	Pacific Hydrogen	1.9	1986	-
Canada	Monterial	Air Liquide	10	1986	Yes
Holland	Rosenburg	APCI	5	1987	Yes
France	Waziers, Lille	Air Liquide	10	1987	Yes
Japan	Minamitane	Japan Liquid Hydrogen	2.2	1987	Yes
USA	Niagara Falls, NY	Praxair	18	1989	Yes
Canada	Magog, Quebec	BOC	15	1989	Yes
Guyana¡	Kouru F.	Air Liquide	5	1990	Yes
Canada	Monterial	BOC	14	1990	Yes
India	Mahendragiri	ISRO	0.3	1992	Yes
Germany	Ingolstadt,,	Linde	4.4	1991	Yes
USA	Pace, FL	APCI	30	1994	Yes
USA	McIntosh, AL	Praxair	24	1995	Yes
China	Beijing	CALT	0.6	1995	Yes
USA	East Chicago, IN	Praxair	30	1997	Yes
Japan	Kimitsu	Air Products	0.3	2003	Yes
India	Saggonda	Andhra Sugars	1.2	2004	Yes
Japan	Osaka	Iwatani	11.3	2006	Yes
Germany	Leuna	Linde	5	2008	Yes
Japan	Tokyo	Iwatani	10	2008	Yes
India	India	Asiatic Oxygen	1.2	-	Yes
USA	California	Stearns-Roger Mfg. Co.	62.5	-	-
USA	New Jersey	Air reduction Sales Co.	6	-	-
USA	Ashtabula, OH	Praxair	-	-	No

* Tons per day. ** Air Products and Chemicals, Inc. ¡ French Guyana.

Table 3. Summary of the published SECs of the industrial hydrogen liquefiers

Hydrogen liquefaction system	**SEC (kWh/kg_{LH2})**	**η_x* (%)**
1- Praxair plant (Drnevich 2003)	12.5-15 kWh/kg_{LH2}	19.3-23.1
2- Ingolstadt Plant (Bracha et al., 1994; Gross et al., 1994)	13.58 kWh/kg_{LH2}	21.2
3- Ingolstadt Plant (Weindorf, Bünger, and Schindler 2003)	13 kWh/kg_{LH2}	22.2
4- Ingolstadt Plant (Quack 2002)	15 kWh/kg_{LH2}	19.3

*Exergy efficiency

Table 4. Technical comparison of the proposed theoretical systems for hydrogen liquefaction

Hydrogen liquefaction system Feed: P (kPa)/T (K)/condition Output: P (kPa)/T (K)/condition	$\dot{m}_f$* (TPD)	y** (%)	SEC $\left(\frac{kWh}{kg_{LH_2}}\right)$	η_x[§] (%)	The configuration according to Figure 5. Remarks
Baker and Shaner (Baker and Shaner, 1978)	250	100	10.85	36.0	LN_2 pre-cooled dual pressure Claude
929/308/25% p − LH_2 100/20.57/97% p − LH_2					H_2 expander isentropic efficiency 0.79, compressor isentropic efficiency 0.79, liquefier max p = 4137 kPa.
Matsuda and Nagamei (WE-NET) (Matsuda and Nagami, 1997)					J-B pre-cooling Claude
5000/300/25% p − LH_2					Expanders isentropic efficiency 0.85 (power recovery efficiency = 0.90), comp. isentropic efficiency 0.80, liquefier max p≈5000 kPa.
100/20.4/> 95% p − LH_2					
Hydrogen J-B pre-cooling Claude	300	100	8.60	46.4	According to the trade-off evaluation of the processes:
Helium J-B pre-cooling Claude	300	100	8.76	45.6	a) refrigerant and heat exchangers costs in the hydrogen case are minimal and costs for compressors and safety are maximum.
Basic neon J-B pre-cooling Claude	300	100	8.65	46.2	b) costs for heat exchangers of helium case are minimum.
Neon with cold pump J-B pre-cooling Claude	300	100	8.49	47.1	c) refrigerant, maintenance and plant cost of neon with cold pump case is maximum and its operation, management of refrigerant, and expanders cost are minimum.

Table 4. (Continued)

Hydrogen liquefaction system Feed: P (kPa)/T (K)/condition Output: P (kPa)/T (K)/condition	$\dot{m}_f$* (TPD)	y** (%)	SEC $\left(\frac{kWh}{kg_{LH_2}}\right)$	η_x[§] (%)	The configuration according to Figure 5. Remarks
Quack (Quack, 2002)					J-B pre-cooling Claude with mixed refrigerant
8000/300/25% $p-LH_2$					MR comprises 20% neon and 80 % helium, expanders isentropic efficiency 0.85-0.90, compressor isentropic efficiency 0.85, liquefier max p≈8000 kPa, heat exchangers minimum temperature approach 2.12 K, increasing the circumferential speed of turbocompressors is reported as the most urgent field of development.
100/20.15/-					
Without pressure drop in calculation	173	100	≈ 7.0	60.7	
With pressure drop in calculation	173	100	7.39	53.8	
Kuzmenko et al., (Kuz'menko et al., 2004)	54	100	12.7	34.6	Helium pre-cooling Claude integrated with a N_2 J-B cycle
1600/300/25% $p-LH_2$					Hydrogen compressor isothermal efficiency 0.6, helium compressor isentropic efficiency 0.53, nitrogen compressor isentropic efficiency 0.53, N_2 cycle expanders isentropic efficiency 0.73-0.75, helium cycle expanders isentropic efficiency 0.85, max pressure of N_2, helium, and H_2 cycles is 28, 25, and 50 bar, respectively.
150/19.9/> 98% $p-LH_2$					
Kuendig et al., (Kuendig et al., 2006)	50	100	4 (Kramer, Huijsmans, and Austgen 2006)	-	LNG bath precooling integrated with a N_2 J-B cycle
2200/300/25% $p-LH_2$					Commercial machinery to realize this concept is presented, LNG is required and precooled feed hydrogen gas and the plant should be located near a seaport.
200/22.8/99.5% $p-LH_2$					

Hydrogen liquefaction system Feed: P (kPa)/T (K)/condition Output: P (kPa)/T (K)/condition	$\dot{m}_f$* (TPD)	y** (%)	SEC $\left(\frac{kWh}{kg_{LH_2}}\right)$	η_x§ (%)	The configuration according to Figure 5. Remarks
Stang et al., (Stang et al., 2006) 2000/300/25% $p-LH_2$ 00/20.3/-	-	100	7.0	60	MR and helium bath pre-cooling Helium-neon mixture is used as the mixed refrigerant, the onset of the inevitable accompaniment of the mixed refrigerant cycles and helium refrigerating cycles.
Notardonato (Notardonato, 2006) 2400/300/25% $p-LH_2$ -/-/-	-	100	-	-	Helium pre-cooling Claude Compressors and expansion turbines isentropic efficiency 0.80, it is shown that the cycle has the maximum efficiency when the hydrogen and helium pressure are set at 2400 kPa and 1200 kPa, respectively.
Shimko and Gardiner (Shimko, 2008) 2100/300/25% $p-LH_2$ 100/20.0/95% $p-LH_2$	50	100	8.73	44.6	Helium pre-cooling Claude Turbines isentropic efficiency 0.83-0.86, wet expander isentropic efficiency 0.90, He and H_2 compessors isentropic efficiency0.8, helium cycle max p≈1500 kPa, hydrogen cycle max p≈2100 kPa, remove of compression load from hydrogen compressors and transfer to helium compressors is the main feature of the cycle.
Staats et al., (Staats et al., 2008) 2100/300/25% $p-LH_2$ 100/20.0/95% $p-LH_2$	50	100	-	39	Helium pre-cooling Claude Turbines isentropic efficiency 0.83-0.86, wet expander isentropic efficiency 0.90, He compressor isentropic efficiency 0.80, H_2 compressor isentropic efficiency 0.60, hydrogen cycle max p≈2100 kPa.

Table 4. (Continued)

Hydrogen liquefaction system Feed: P (kPa)/T (K)/condition Output: P (kPa)/T (K)/condition	$\dot{m}_f$* (TPD)	y** (%)	SEC $\left(\frac{kWh}{kg_{LH_2}}\right)$	η_x§ (%)	The configuration according to Figure 5. Remarks
Valenti (Valenti and Macchi, 2008) 6000/300/25% $p-LH_2$ 150/20/-	864	100	5.04	47.7	J-B pre-cooling Claude with helium as the refrigerant The polytrophic efficiency of helium expanders 0.88-0.93, helium compressors polytrophic efficiency 0.92, hydrogen expander polytrophic efficiency 0.85, J-B cycle max p = 4000 kPa, heat exchangers minimum temperature approach is 2.30-3.15 K, it is reported that helium compressors and heat exchangers have the highest exergy loss.
Krasae-in et al., (Krasae-In et al., 2010b) 2100/298.15/25% $p-LH_2$ 130/19.45/95% $p-LH_2$	100	100	5.35	54.0	MR and J-B pre-cooling Claude Mixed refrigerant with ten components, compressors and turbines isentropic efficiency 0.80, heat exchangers minimum temperature approach 1-2 K J-B cycles max p = 4000 kPa, MR cycle max p = 1800 kPa.
Krasae-in (Krasae-In, 2014) 2100/298.15/25% $p-LH_2$ 130/19.45/95% $p-LH_2$	100	100	5.91	48.9	MR and J-B pre-cooling Claude Five-component mixture consists of 4% hydrogen, 18% nitrogen, 24% methane, 28% ethane, and 26% butane (by mole), J-B cycles max p=4000 kPa, MR cycle max p = 1800 kPa, it is concluded that pressure drops in the heat exchangers do not have a significant impact on the total power consumption.

Hydrogen liquefaction system Feed: P (kPa)/T (K)/condition Output: P (kPa)/T (K)/condition	$\dot{m}_f$* (TPD)	y** (%)	SEC $\left(\frac{kWh}{kg_{LH_2}}\right)$	η_x§ (%)	The configuration according to Figure 5. Remarks
Yuksel et al (Yuksel et al., 2017)	50 (Staats, Smith, and Brisson 2008)	100	-	57.13	Helium pre-cooling Claude
2100/298.15/25% $p-LH_2$ (Staats, Smith, and Brisson 2008)					Helium and H_2 compressors isentropic efficiency 0.80, helium expanders efficiency is 0.85, it is concluded that the largest exergy destruction rates occur in helium turbo expanders.
202/20.55/99% $p-LH_2$					
Sadaghiani and Mehrpooya (Sadaghiani and Mehrpooya, 2017)	≈ 300	100	4.36	55.47	MR and J-B pre-cooling Claude
2100/298.15/25% $p-LH_2$					Mixed refrigerant with nine components for the pre-cooling cycle, a mixed refrigerant with three components for the cryogenic J-B cycle, efficiency of compressors and expanders is supposed to be 0.90 and 0.85, respectively, heat exchangers minimum temperature approach is 1.4-2.2K. J-B cycles max p = 1000 kPa. MR cycle max p = 1600 kPa.
130/20.15/95% $p-LH_2$					

Table 4. (Continued)

Hydrogen liquefaction system Feed: P (kPa)/T (K)/condition Output: P (kPa)/T (K)/condition	$\dot{m}_f$* (TPD)	y** (%)	SEC $\left(\frac{kWh}{kg_{LH_2}}\right)$	η_x[§] (%)	The configuration according to Figure 5. Remarks
Asadnia and Mehrpooya (Asadnia and Mehrpooya, 2017)	≈ 100	100	7.69	39.5	MR and J-B pre-cooling Claude
$2100/298.15/25\%\ p-LH_2$ $130/20.95/95\%\ p-LH_2$					Mixed refrigerant with 11 components for the pre-cooling cycle, a mixed refrigerant with two components for cryogenic J-B cycles, isentropic efficiency of compressors and expanders is supposed to be 0.8, heat exchangers minimum temperature approach is 1-2K. J-B cycles max p = 3000 kPa, MR cycle max p = 1805 kPa, it is reported that heat exchangers and compressors, the most expensive equipment, including 58% of the total exergy losses.
Hammad and Dincer (Hammad and Dincer, 2018)	5.20	25	3.66	11.58	LN_2 pre-cooled Claude
$1000/298.15/25\%\ p-LH_2$					Pressure drops and heat losses in pipes are neglected, a heat loss of 20% is considered in heat exchangers, it is showed that one of the expanders has the lowest exergy efficiency.

* Capacity (Tones per day).

** Liquid yield ($y = \dot{m}_f/\dot{m}$).

§ Exergy efficiency.

Table 5. Typical data for a large-scale conceptual plant for the liquefaction of hydrogen (Asadnia and Mehrpooya, 2017)

Technical parameter	Feed hydrogen	Produced LH_2
Pressure (bar)	21	1.3
Temperature (K)	≈ 300	≈ 21
Mass flow rate (kg/s)	1.157	1.157
$p - LH_2$ (%)	25	95
Lowest pressure (bar)	1.3	
Highest pressure (bar)	21	
H_2 Expander exit temperature (K)	21	
Temperature approach of heat exchangers (K)	1-2	
Isentropic efficiency of the expanders (%)	80	
Isentropic efficiency of compressors (%)	80	
Work of liquefaction (kWh/kg_{LH2})	7.69	
Exergy efficiency (%)	39.5	

Hydrogen Liquefying Using Combined Hybrid Plants

A hybrid process is a combination of one of the conventional hydrogen liquefying processes and renewable energy sources. The development of such a process has started recently (for example, Muradov and Veziroğlu, 2008). The implementation of the geothermal energy systems for required electricity as well as the absorption refrigeration systems for required heat (ARS) are discussed by Kanoglu et al., (2007), Ratlamwala et al., (2012), Gadalla et al., (2013). The refrigeration capacity of the ARS is used to precool feeds or any other streams through liquefying process (for example, Yilmaz et al., 2012; Kanoglu et al., 2016a). In some cases, the hot water from the geothermal system is used to produce hydrogen through a high-temperature steam electrolysis (Lipman, 2011). The combination of hydrogen liquefaction processes and systems such as e.g., solar photovoltaic/thermal systems (Ratlamwala et al., 2012), nuclear-based energy systems (Ozcan and Dince, 2016), and a low-temperature auto-cascade solar Rankine cycle (ALSRC) system (Bao et al., 2011) can be considered as hybrid configurations as well. Figure 9 generalizes the described integrated hybrid system. The different auxiliary systems may be used to supply the required power or to precool different streams of the Claude cycle.

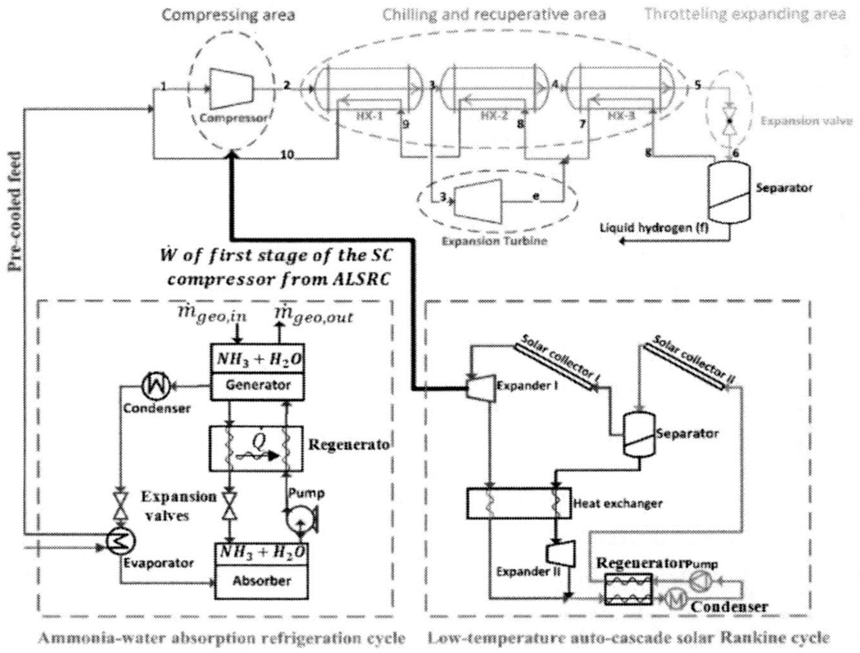

Figure 9. Typical PFD of an integrated hybrid system (Asadnia and Mehrpooya, 2017). ARS/ALSRC/SC combined systems (Bao et al., 2011; Kanoglu et al., 2016b).

HYDROGEN LIQUEFACTION SYSTEMS ANALYSIS

Figure 10 shows the main framework of the gas liquefaction cycles. As it can be seen, a purification is the first step. After the purification, creating a non-ideality of the gas is a mandatory stage. It is usually done by increasing the gas pressure/the decreasing temperature. The next significant step is related to the gas inversion temperature. If it is lower than the ambient temperature, the gas should be precooled. However, other gases pass the bypass route to get to the destination where the liquefying takes place. The bypass method was the methodology used for gas liquefying before 1895. Finally, the gas liquefaction process is completed with an expansion unit.

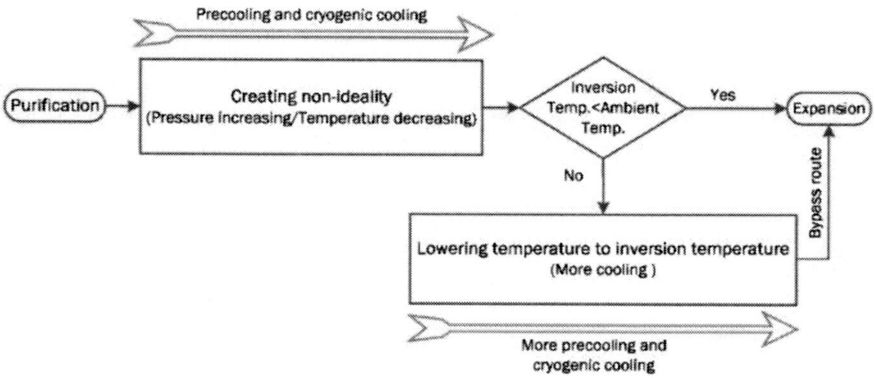

Figure 10. Overview of conventional gas liquefaction methodology (from Asadnia and Mehrpooya, 2017).

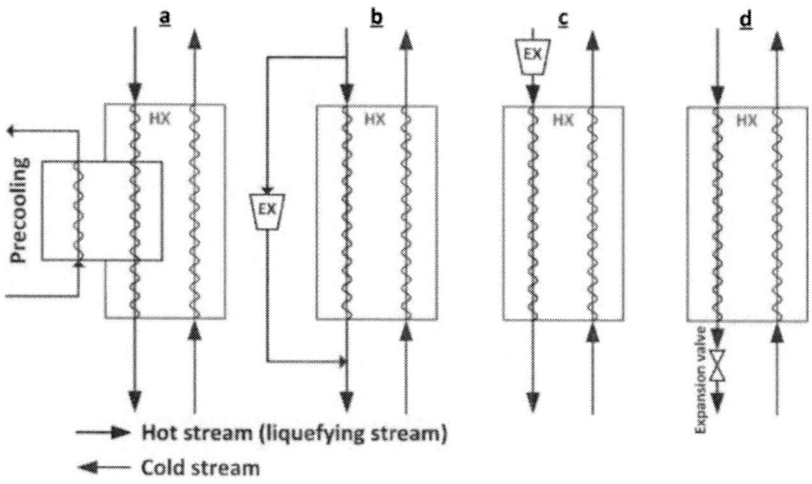

Figure 11. The cooling methodology of the liquefied gas (from Asadnia and Mehrpooya, 2017): a) pre-cooling with a cryogenic liquid bath or auxiliary refrigeration systems, b) parallel isentropic expansion, and c) serial isentropic expansion.

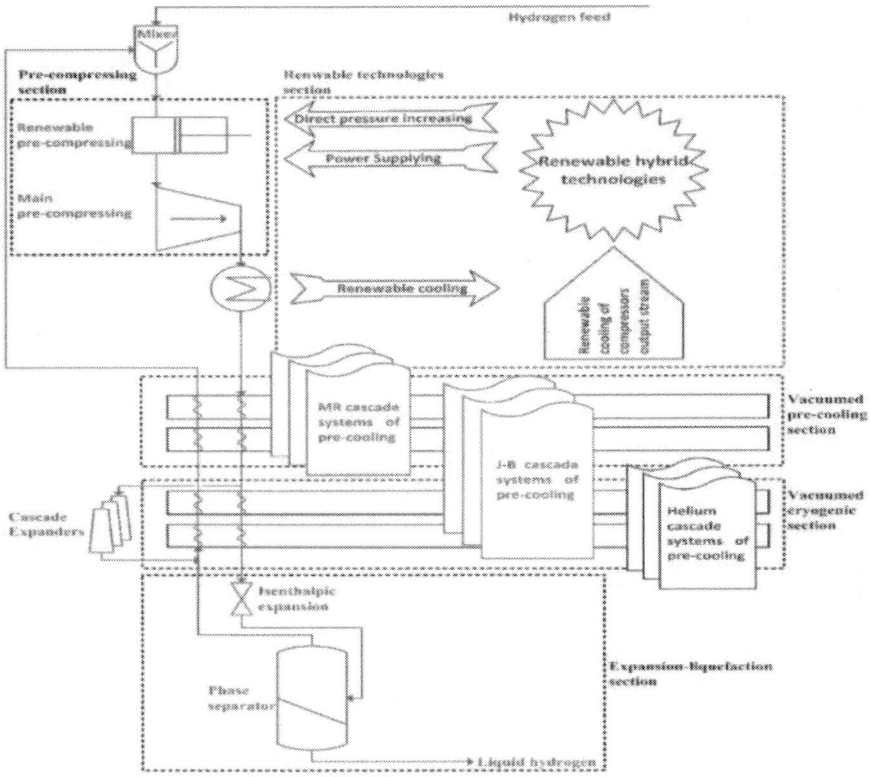

Figure 12. The overall configuration of LHL systems emphasizing the status of renewable energy technologies in the context of hydrogen liquefaction (from Asadnia and Mehrpooya, 2017).

Figure 11 shows conventional technologies used to decrease the gas temperature. Figure 12 depicts a typical block flow diagram of the overall configuration of LHL systems. Figure 13. A novel classification of hydrogen liquefaction systems in terms of heat exchange and expansion methods (from Asadnia and Mehrpooya, 2017).

Table 6 lists all of the specific energy consumptions that are noted in the publications, corresponding to their citation in this chapter.

Table 6. SEC of hydrogen liquefaction cycles

The configuration of the hydrogen liquefaction cycle	SEC (kWh/kg_{LH_2})
Simple Claude (Asadnia and Mehrpooya, 2017)	22.10
Pre-cooled ideal Linde-Hampson cycle (Arend and Chelton, 1964)	16.24
Pre-cooled Linde-Hampson cycle (Nandi and Sarangi, 1993)	68.10
Pre-cooled single pressure Claude cycle (Nandi and Sarangi, 1993)	29.90
Pre-cooled ideal dual pressure Linde-Hampson cycle (Arend and Chelton, 1964)	12.12
Pre-cooled ideal dual pressure Claude cycle (Arend and Chelton, 1964)	6.66
Pre-cooled dual pressure Claude cycle (Arend and Chelton, 1964)	12.26
Helium-refrigerated hydrogen liquefaction system (Nandi and Sarangi, 1993)	44.80
WE-NET project in Japan (Ohira, 2004)	7.90
Praxair plants' medium SEC in the US (Drnevich, 2003)	13.75
Linde AG large-scale plant in Ingolstadt (Bracha et al., 1994; Gross et al., 1994)	13.58
Linde AG large-scale plant in Ingolstadt (Weindorf et al., 2003)	13.00
Linde AG large-scale plant in Ingolstadt Germany (Quack, 2002)	15.00
Baker system's medium SEC (Baker and Shaner, 1978)	10.42
Matsuda (WE-NET project) (Matsuda and Nagami, 1997)	8.50
Large-scale conceptual plant by Quack (Quack, 2002)	7.00
Kuzmenko helium refrigeration cycle, pre-cooled with LN_2 (Kuz'menko et al., 2004)	12.70
Stang MR cycle (Stang et al., 2006)	7.00
Shimko helium refrigeration cycle (Shimko, 2008)	8.70
Four helium Joule–Brayton cascade cycle by Valenti Valenti and Macchi, 2008)	5.76
Large-scale MR system by Krasae-in in 2010 (Krasae-in et al., 2010b)	5.35
Large-scale MR system by Krasae-in in 2014 (Krasae-in, 2014)	5.91
Large-scale LH_2 production conceptual plants by Sadaghiani (Sadaghiani and Mehrpooya, 2017)	4.41
Mixed refrigerant LHL combined system by Asadnia (Asadnia and Mehrpooya, 2017)	7.69
The thermodynamically ideal liquefaction system (Krasae-in et al., 2010)	2.89
The thermodynamically ideal liquefaction system Arend and Chelton, 1964; Barron, 1972; Peschka, 2012;)	3.92

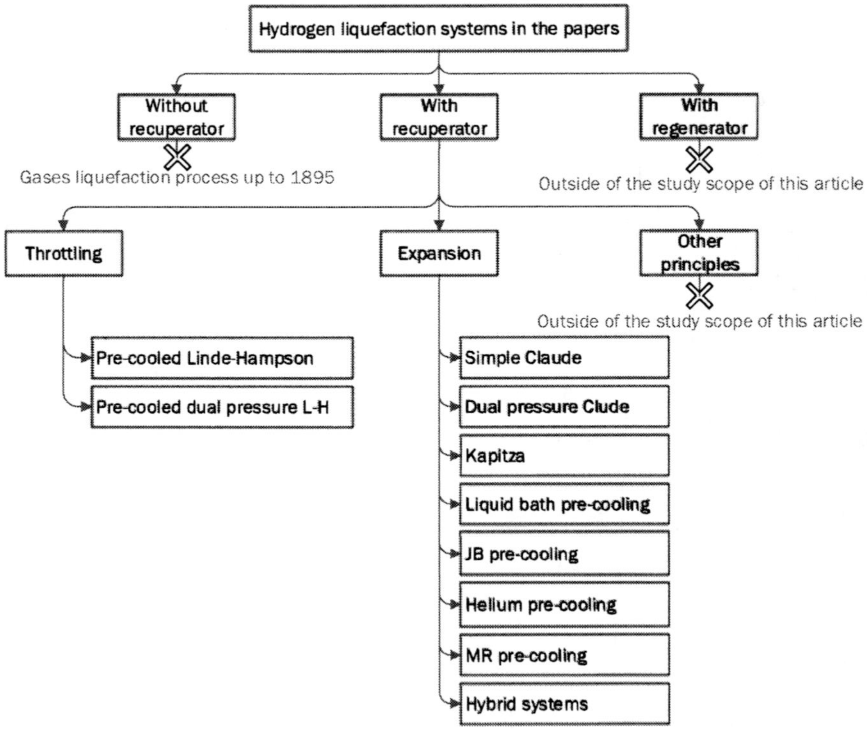

Figure 13. A novel classification of hydrogen liquefaction systems in terms of heat exchange and expansion methods (from Asadnia and Mehrpooya, 2017).

SUMMARY AND CONCLUSION

The different configurations of the hydrogen liquefaction process are discussed. The different technologies for hydrogen liquefying are reviewed and categorized in terms of the expansion process and heat recuperation method. Four main areas were determined: (a) the substitution of throttling valves with turbines; (b) multi-level refrigeration or cascading systems; (c) the use of the mixed refrigerants, and (d) the implementation of renewable energy sources to the hydrogen liquefying systems. It is predicted that the SEC of the conceptual proposed plant will not be significantly reduces in the near future.

REFERENCES

Al-Hallaj, Said, Stephen Wilke, and Ben Schweitzer. 2017. "Energy Storage Systems for Smart Grid Applications." In *Water, Energy & Food Sustainability in the Middle East*, 161–92. Springer.

Ansarinasab, Hojat, Mehdi Mehrpooya, and Amin Mohammadi. 2017. "Advanced Exergy and Exergoeconomic Analyses of a Hydrogen Liquefaction Plant Equipped with Mixed Refrigerant System." *Journal of Cleaner Production* 144. Elsevier Ltd: 248–59. doi:10.1016/j.jclepro.2017.01.014.

Arend, P. C. Wander, and D. B. Chelton. 2013. "3B2. Large-Scale Hydrogen Liquefaction Facilities." *Technology and Uses of Liquid Hydrogen*. Elsevier, 79.

Asadnia, M., and M. Mehrpooya. 2017. "A Novel Hydrogen Liquefaction Process Configuration with Combined Mixed Refrigerant Systems." *International Journal of Hydrogen Energy* 42 (23). doi:10.1016/j.ijhydene.2017.04.260.

Asadnia, Majid, and Mehdi Mehrpooya. 2017. "Large-Scale Liquid Hydrogen Production Methods and Approaches: A Review." *Applied Energy*.

Baker, C. R., and R. L. Shaner. 1978. "A Study of the Efficiency of Hydrogen Liquefaction." *International Journal of Hydrogen Energy* 3 (3): 321–34. doi:10.1016/0360-3199(78)90037-X.

Bao, J. J., L. Zhao, and W. Z. Zhang. 2011. "A Novel Auto-Cascade Low-Temperature Solar Rankine Cycle System for Power Generation." *Solar Energy* 85 (11). Elsevier Ltd: 2710–19. doi:10.1016/j.solener.2011.08.015.

Barron, R F. 1972. "Liquefaction Cycles for Cryogens." In *Advances in Cryogenic Engineering*, 20–36. Springer.

Belyakov, V. V, and B. D Krakovskii. 2002. "Low-Capacity Hydrogen Liquefier with a Helium Cycle." *Chemical and Petroleum ...* 38 (3): 26–28. http://www.springerlink.com/index/m648852jp74650k3.pdf.

Berstad, David, Jacob Stang, and Petter Nekså. 2009. "*A Future Energy Chain Based on Liquefied Hydrogen*," SINTEF Energy Research, 1–23.

Berstad, David, Harald T. Walnum, and P. Nekså. 2013. "Integrated Design for Demonstration of Efficient Liquefaction of Hydrogen (IDEALHY)." *Fuel Cells and Hydrogen Joint Undertaking (FCH JU).*

Bliesner, Ronald M. 2013. "*Parahydrogen-Orthohydrogen Conversion for Boil-Off Reduction From Space Stage.*" Washington State University.

Bracha, M., G. Lorenz, A. Patzelt, and M. Wanner. 1994. "Large-Scale Hydrogen Liquefaction in Germany." *International Journal of Hydrogen Energy* 19 (1): 53–59. doi:10.1016/0360-3199(94)90177-5.

Brickwedde, F. G., R. B. Scott, and H. S. Taylor. 1935. "The Difference in Vapor Pressures of Ortho-and Paradeuterium." *J. Res. Natl. Bur. Std* 15: 463–75.

Cardella, U., L. Decker, and H. Klein. 2017a. "Economically Viable Large-Scale Hydrogen Liquefaction." In *IOP Conference Series: Materials Science and Engineering*, 171:12013. IOP Publishing.

Cardella, U., L. Decker, and H. Klein. 2017b. "*ScienceDirect Roadmap to Economically Viable Hydrogen Liquefaction.*" Elsevier Ltd, 1–10.

Choudhury, Balaji Kumar. 2009. "*Process Design of Turboexpander Based Nitrogen Liquefier.*" MTech thesis.

Davies, Mansel. 1989. "William Hampson (1854–1926): A Note." *The British Journal for the History of Science* 22 (1). Cambridge University Press: 63–73.

Denton, W. H., and C. M. Nicholls. 1964. *Technology and Uses of Liquid Hydrogen*. Edited by Scott R. B. Pergamon Press. doi:10.1016/B978-0-08-010779-0.50004-5.

Dewar, James. 1898. "Liquid Hydrogen." *Science* 8 (183): 3–6.

Drnevich, Raymond. 2003. "*Hydrogen Delivery, Liquefaction & Compression. Strategic Initiatives for Hydrogen Delivery Workshop.*" New York.

Edeskuty, F. J. 1964. "Liquid Hydrogen as a Coolant/propellant for Nuclear Rockets." In *Technology and Uses of Liquid Hydrogen*, 181–94.

Eriksson, E. L. V., and E. MacA Gray. 2017. "Optimization and Integration of Hybrid Renewable Energy Hydrogen Fuel Cell Energy systems–A Critical Review." *Applied Energy* 202. Elsevier: 348–64.

Flynn, Thomas M. 2005. *Cryogenic Engineering: Revised and Expanded, Second Edition*. Dekker, NY.

Fradkov, A. B., and V. F. Troitskii. 1965. "Liquefier with Two-Stage Conversion to Obtain 98 percent Parahydrogen." *Cryogenics* 5 (3). Elsevier: 136–37.

Gadalla, Mohamed A., Tahir Abdul Hussain Ratlamwala, Ibrahim Dincer, and Mehmet Kanoglu. 2013. "Performance Assessment of an Integrated Absorption Cooling–hydrogen Liquefaction System Using Geothermal Energy." *International Journal of Exergy* 12 (2). Inderscience Publishers: 205–25.

Giauque, W. F., and H. L. Johnston. 1928. "Symmetrical and Antisymmetrical Hydrogen and the Third Law of Thermodynamics. Thermal Equilibrium and the Triple Point Pressure." *Journal of the American Chemical Society* 50 (12). ACS Publications: 3221–28.

Gross, Reinhard, Wolfgang Otto, Adolf Patzelt, and Manfred Wanner. 1994. "Liquid Hydrogen for Europe–the Linde Plant at Ingolstadt." *Linde–Reports on Science and Technology* 54: 37–43.

Gursu, S, S. A. Sherif, T. N. Veziroglu, and J. W. Sheffield. 1993. "Analysis and Optimization of Thermal Stratification and Self-Pressurization Effects in Liquid Hydrogen Storage systems—Part 1: Model Development." *Journal of Energy Resources Technology* 115 (3). American Society of Mechanical Engineers: 221–27.

Hammad, Anwar, and Ibrahim Dincer. 2018. "Analysis and Assessment of an Advanced Hydrogen Liquefaction System." *International Journal of Hydrogen Energy* 43 (2). Elsevier: 1139–51.

Haselden, Geoffrey G. 1971. *Cryogenic Fundamentals*. New York, Academic Press Inc.

Hemmati, Reza, and Hedayat Saboori. 2016. "Emergence of Hybrid Energy Storage Systems in Renewable Energy and Transport applications–A Review." *Renewable and Sustainable Energy Reviews* 65. Elsevier: 11–23.

Hernandez, H. P., and B. W. Birmingham. 1964. "Liquid Hydrogen Bubble Chambers." In *Technology and Uses of Liquid Hydrogen*, 228–84.

Jensen, J. E., Rex George Stewart, W. A. Tuttle, and H. Brechna. 1980. *Brookhaven National Laboratory Selected Cryogenic Data Notebook: Sections I-IX.* Vol. 1. Brookhaven National Laboratory.

Kalanidhi, A. 1988. "Boil-off in Long-Term Stored Liquid Hydrogen." *International Journal of Hydrogen Energy* 13 (5). Elsevier: 311–13.

Kanoglu, Mehmet, Ibrahim Dincer, and Marc A. Rosen. 2007. "Geothermal Energy Use in Hydrogen Liquefaction." *International Journal of Hydrogen Energy* 32 (17): 4250–57. doi:10.1016/j.ijhydene.2007.06.006.

Kanoglu, Mehmet, Ceyhun Yilmaz, and Aysegul Abusoglu. 2016a. "Geothermal Energy Use in Absorption Precooling for Claude Hydrogen Liquefaction Cycle." *International Journal of Hydrogen Energy*. Elsevier Ltd, 1–16. doi:10.1016/j.ijhydene.2016.04.068.

———. 2016b. "ScienceDirect Geothermal Energy Use in Absorption Precooling for Claude Hydrogen Liquefaction Cycle." *International Journal of Hydrogen Energy*. Elsevier Ltd. doi:10.1016/j.ijhydene.2016.04.068.

Kapitza, P. 1934. "The Liquefaction of Helium by an Adiabatic Method." *Proceedings of the Royal Society of London. Series A, Mathematical and Physical Sciences* 147 (860). JSTOR: 189–211.

Kerry, Frank. 2006. *Industrial Gas Handbook Gas Separation. Gas Separation and Purification.*

Kleemenko, Aleksander Petrovich. 1959. "One Flow Cascade Cycle." In *Proc. of the 10th International Congress of Refrigeration*, 1:34–39.

Kramer, Gert Jan, Joep Huijsmans, and Dave Austgen. 2006. "Clean and Green Hydrogen." *WHEC* 16: 1–9.

Krasae-In, Songwut. 2014. "Optimal Operation of a Large-Scale Liquid Hydrogen Plant Utilizing Mixed Fluid Refrigeration System." *International Journal of Hydrogen Energy* 39 (13). Elsevier Ltd: 7015–29. doi:10.1016/j.ijhydene.2014.02.046.

Krasae-In, Songwut, Arne M. Bredesen, Jacob H. Stang, and Petter Neksa. 2011. "Simulation and Experiment of a Hydrogen Liquefaction Test Rig Using a Multi-Component Refrigerant Refrigeration System."

International Journal of Hydrogen Energy 36 (1). Elsevier Ltd: 907–19. doi:10.1016/j.ijhydene.2010.09.005.

Krasae-in, Songwut, Jacob H. Stang, and Petter Neksa. 2010. "Development of Large-Scale Hydrogen Liquefaction Processes from 1898 to 2009." *International Journal of Hydrogen Energy* 35 (10). Elsevier Ltd: 4524–33. doi:10.1016/j.ijhydene.2010.02.109.

Krasae-In, Songwut, Jacob H. Stang, and Petter Neksa. 2010a. "Exergy Analysis on the Simulation of a Small-Scale Hydrogen Liquefaction Test Rig with a Multi-Component Refrigerant Refrigeration System." *International Journal of Hydrogen Energy* 35 (15). Elsevier Ltd: 8030–42. doi:10.1016/j.ijhydene.2010.05.049.

Krasae-In, Songwut, Jacob H. Stang, and Petter Neksa. 2010b. "Simulation on a Proposed Large-Scale Liquid Hydrogen Plant Using a Multi-Component Refrigerant Refrigeration System." *International Journal of Hydrogen Energy* 35 (22). Elsevier Ltd: 12531–44. doi:10.1016/j.ijhydene.2010.08.062.

Kubbinga, Henk. 2010. "A Tribute to Wróblewski and Olszewski." *Europhysics News* 41 (4). EDP Sciences: 21–24.

Kuendig, Andres, Karl Lorhlein, Gert Jan Kramer, and Joep Huijsmans. 2006. "Large Scale Hydrogen Liquefaction in Combination with LNG Re-Gasification." In *Proceedings of the 16th World Hydrogen Energy Conference 2006, Lyon, France*, 3326–33.

Kuz'menko, I. F., I. M. Morkovkin, and E. I. Gurov. 2004. "Concept of Building Medium-Capacity Hydrogen Liquefiers with Helium." *Chemical and Petroleum Engineering* 40 (1–2): 94–98.

Li, Yongliang, Haisheng Chen, and Yulong Ding. 2010. "Fundamentals and Applications of Cryogen as a Thermal Energy Carrier: A Critical Assessment." *International Journal of Thermal Sciences* 49 (6). Elsevier Masson SAS: 941–49. doi:10.1016/j.ijthermalsci.2009.12.012.

Lipman, Timothy. 2011. "An Overview of Hydrogen Production and Storage Systems with Renewable Hydrogen Case Studies." *Clean Energy States Alliance*, 32. doi:10.1017/CBO9781107415324.004.

Matsuda, H, and M Nagami. 1997. *"Study of Large Hydrogen Liquefaction Process."* [*Hydrogen Energy*] 8(3). [Korean Hydrogen & New Energy Society]: 175.

Maytal, Ben-Zion, and John M. Pfotenhauer. 2012. *Miniature Joule-Thomson Cryocooling: Principles and Practice*. Springer Science & Business Media.

McCarty, RD, J Hord, and H.M. Roder. 1981. "*NBS Monograph 168 Selected Properties of Hydrogen (Engineering Design Data)*." http://www.osti.gov/energycitations/product.biblio.jsp?osti_id=5747344.

Mehrpooya, Mehdi, Mehdi Omidi, and Ali Vatani. 2016. "Novel Mixed Fluid Cascade Natural Gas Liquefaction Process Configuration Using Absorption Refrigeration System." *Applied Thermal Engineering* 98. Elsevier Ltd: 591–604. doi:10.1016/j.applthermaleng.2015.12.032.

Mukhopadhyay, Mamata. 2010. *Fundamentals of Cryogenic Engineering*. PHI Learning Pvt. Ltd.

Muradov, Nazim Z., and T. Nejat Veziroğlu. 2008. "'Green' Path from Fossil-Based to Hydrogen Economy: An Overview of Carbon-Neutral Technologies." *International Journal of Hydrogen Energy* 33 (23). Elsevier: 6804–39.

Nandi, T. K., and S. Sarangi. 1993. "Performance and Optimization of Hydrogen Liquefaction Cycles." *International Journal of Hydrogen Energy* 18 (2): 131–39. doi:10.1016/0360-3199(93)90199-K.

Nogal, Frank Del, Jin-Kuk Kim, Simon Perry, and Robin Smith. 2008. "Optimal Design of Mixed Refrigerant Cycles." *Industrial & Engineering Chemistry Research* 47 (22): 8724–40. doi: 10.1021/ie800515u.

Notardonato, William Usilton. 2006. *Analysis and Testing of an Integrated Refrigeration and Storage System for Liquid Hydrogen Zero Boil-Off, Liquefaction, and Densification*. Vol. 68.

Ohira, K. 2004. "A Summary of Liquid Hydrogen and Cryogenic Technologies in Japan's WE- NET Project." In *AIP Conference Proceedings*, 710:27–34. AIP.

Ohlig, K., and L. Decker. 2014. "The Latest Developments and Outlook for Hydrogen Liquefaction Technology." *20th World Hydrogen Energy Conference, WHEC 2014* 1: 639–46. doi:10.1063/1.4860858.

Ozcan, Hasan, and Ibrahim Dince. 2016. "Thermodynamic Modeling of a Nuclear Energy Based Integrated System for Hydrogen Production and Liquefaction." *Computers & Chemical Engineering* 90. Elsevier Ltd: 234–46. doi:10.1016/j.compchemeng.2016.04.015.

Peschka, Walter. 2012. *Liquid Hydrogen: Fuel of the Future*. Springer Science & Business Media.

Quack, H. 2002. "Conceptual Design of a High Efficiency Large Capacity Hydrogen Liquefier." *AIP Conference Proceedings* 613 (1): 255–63. doi:10.1063/1.1472029.

Quack, Hans. 2002. "Die Schlusselrolle Der Kryotechnik in Der Wasserstoff-Energiewirtschaft." [The Key Role of Cryogenics in the Hydrogen Energy Industry] *Ki Luft und Kaeltetechnik*, no. 3. CF Mueller Verlag, Huethig GmbH: 157–61.

Rafalska- Łasocha, Alicja. 2015. "Karol Olszewski's 100th Anniversary of Death and His Contribution to Cryogenics (in Polish)." *Prace Komisji Historii Nauki PAU* 14 (1). Directory of Open Access Journals: 335–44.

Ratlamwala, T. A. H., I. Dincer, M. A. Gadalla, and M. Kanoglu. 2012. "Thermodynamic Analysis of a New Renewable Energy Based Hybrid System for Hydrogen Liquefaction." *International Journal of Hydrogen Energy* 37 (23): 18108–17. doi:10.1016/j.ijhydene.2012.09.036.

Ratlamwala, T. A. H., I. Dincer, and M. A. Gadalla. 2012. "Thermodynamic Analysis of a Novel Integrated Geothermal Based Power Generation-Quadruple Effect Absorption Cooling-Hydrogen Liquefaction System." *International Journal of Hydrogen Energy* 37 (7). Elsevier Ltd: 5840–49. doi:10.1016/j.ijhydene.2011.12.119.

Sadaghiani, Mirhadi S., and Mehdi Mehrpooya. 2017. "Introducing and Energy Analysis of a Novel Cryogenic Hydrogen Liquefaction Process Configuration." *International Journal of Hydrogen Energy* 42 (9). Elsevier Ltd: 6033–50. doi:10.1016/j.ijhydene.2017.01.136.

Schmauch, George E, and Alan H Singleton. 1964. "Technical Aspects of Ortho-Parahydrogen Conversion." *Industrial & Engineering Chemistry* 56 (5). ACS Publications: 20–31.

Scurlock, R. G. 1994. "100 Years after Dewar's Dewar: Cryogenics Today." In *Proceedings of the Royal Institution of Great Britain*, 65:145. Oxford University Press, USA.

Sharma, Sunita, and Sib Krishna Ghoshal. 2015. "Hydrogen the Future Transportation Fuel: From Production to Applications." *Renewable and Sustainable Energy Reviews* 43. Elsevier: 1151–58. doi:10.1016/j.rser. 2014.11.093.

Sherif, S. A., N. Zeytinoglu, and T. N. Veziroglu. 1997. "Liquid Hydrogen: Potential, Problems and a Proposed Research Program." *International Journal of Hydrogen Energy* 22 (7): 683–88. doi:10.1016/S0360-3199(96)00201-7.

Shimko, Martin A. 2008. "Innovative Hydrogen Liquefaction Cycle." *FY 2008 Annual Progress Report.*

Sloane. T. 1920. *Liquid Air and the Liquefaction of Gases*. London: The Norman W. Henley.

Staats, W. L., J. L. Smith, and J. G. Brisson. 2008. "*Analysis of a Supercritical Hydrogen Liquefaction Cycle*," 721–30.

Stang, Jacob, Petter Nekså, and Einar Brendeng. 2006. "On the Design of an Efficient Hydrogen Liquefaction Process." In *WHEC*, 1–6. Lyon France.

Stroman, Richard O., Michael W. Schuette, Karen Swider-Lyons, Joseph A. Rodgers, and Daniel J. Edwards. 2014. "Liquid Hydrogen Fuel System Design and Demonstration in a Small Long Endurance Air Vehicle." *International Journal of Hydrogen Energy* 39 (21): 11279–90. doi: 10.1016/j.ijhydene.2014.05.065.

Travers, Morris W. 1899. "The Liquefaction of Hydrogen." *Proceedings of the Physical Society of London* 17 (1). IOP Publishing: 561.

Trevisani, L., M. Fabbri, F. Negrini, and P. L. Ribani. 2007. "*Advanced Energy Recovery Systems from Liquid Hydrogen*" 48: 146–54. doi:10.1016/j.enconman.2006.05.002.

Tzimas, Evangelos, C. Filiou, S. D. Peteves, and J. B. Veyret. 2003. "Hydrogen Storage: State-of-the-Art and Future Perspective." *EU Commission, JRC Petten, EUR 20995EN.*

Uyar, Tanay Sıdkı, and Doğancan Beşikci. 2017. "Integration of Hydrogen Energy Systems into Renewable Energy Systems for Better Design of 100% Renewable Energy Communities." *International Journal of Hydrogen Energy* 42 (4). Elsevier: 2453–56.

Valenti, Gianluca, and Ennio Macchi. 2008. "Proposal of an Innovative, High-Efficiency, Large-Scale Hydrogen Liquefier." *International Journal of Hydrogen Energy* 33 (12): 3116–21. doi:10.1016/j.ijhydene.2008.03.044.

Vander Arend, P C, and D B Chelton. 1964. "The liquefaction of hydrogen 3a. Basic principles." *Technology and Uses of Liquid Hydrogen*. Pergamon, 38.

Venkatarathnam, Gadhiraju, and Klaus D Timmerhaus. 2008. *Cryogenic Mixed Refrigerant Processes*. Edited by Klaus D. Timmerhaus. Vol. 100. New York: Springer.

Walker, Graham. 1983. *Cryocoolers. Part 1: Fundamentals*. Edited by Klaus D. Timmerhaus, F. Clark Clark, J. L. Olsen, and Kurt Mendelssohn. New York: Springer Science & Business Media. doi:10.1007/978-1-4899-5286-8.

Walker, Graham. 1983. 2012. *Cryocoolers: Part 2: Applications*. Springer Science & Business Media.

Walnum H. T., Berstad D., Drescher M., Petter Neksa, H. Quack, and Ch. Haberstroh. 2012. "Principles for the Liquefaction of Hydrogen with Emphasis on Precooling Processes." In *12th Cryogenics 2012*, 8. http://www.idealhy.eu/uploads/documents/IDEALHY_Cryogenics_2012_Precooling.pdf.

Wang, Xin, Jun Yang, Lei Chen, and Jifeng He. 2017. "Application of Liquid Hydrogen with SMES for Efficient Use of Renewable Energy in the Energy Internet." *Energies* 10 (2): 185. doi:10.3390/en10020185.

Webb, F. J. 1964. "The Use of Liquid Hydrogen for the Production of Cold Neutrons inside a Nuclear Reactor." In *Technology and Uses of Liquid Hydrogen*, 195–227.

Weindorf, Werner, Ulrich Bünger, and Jörg Schindler. 2003. "*Comments on the Paper by Baldur Eliasson and Ulf Bossel 'The Future of the Hydrogen Economy: Bright or Bleak?'*" LB-Systemtechnik GmbH, Ottobrunn.

West, J. E. 2003. "*The Economics of Small to Medium Liquid Hydrogen Facilities.*"

Windmeier, Christoph, and Randall F Barron. 2000. "Cryogenic Technology." *Ullmann's Encyclopedia of Industrial Chemistry*. Wiley Online Library.

Wolf, Joachim. 2002. "Liquid Hydrogen Technology for Vehicles." *MRS Bulletin* 27 (9): 684–87. doi:10.1557/mrs2002.222.

Won, Wangyun, Hweeung Kwon, Jee-Hoon Han, and Jiyong Kim. 2017. "Design and Operation of Renewable Energy Sources Based Hydrogen Supply System: Technology Integration and Optimization." *Renewable Energy* 103. Elsevier Ltd: 226–38. doi:10.1016/j.renene.2016.11.038.

Yang, Christopher, and Joan Ogden. 2007. "Determining the Lowest-Cost Hydrogen Delivery Mode." *International Journal of Hydrogen Energy* 32 (2): 268–86. doi:10.1016/j.ijhydene.2006.05.009.

Yilmaz, Ceyhun, Mehmet Kanoglu, Ali Bolatturk, and Mohamed Gadalla. 2012. "Economics of Hydrogen Production and Liquefaction by Geothermal Energy." *International Journal of Hydrogen Energy* 37 (2). Elsevier Ltd: 2058–69. doi:10.1016/j.ijhydene.2011.06.037.

Yuksel, Yunus Emre, Murat Ozturk, and Ibrahim Dincer. 2017. "Analysis and Assessment of a Novel Hydrogen Liquefaction Process." *International Journal of Hydrogen Energy*. Elsevier Ltd, 1–10. doi:10.1016/j.ijhydene.2017.03.064.

Zhang, Jinsong, Timothy S Fisher, P Veeraraghavan Ramachandran, Jay P Gore, and Issam Mudawar. 2005. "A Review of Heat Transfer Issues in Hydrogen Storage Technologies." *Journal of Heat Transfer* 127 (12). American Society of Mechanical Engineers: 1391–99.

BIOGRAPHICAL SKETCH

Mehdi Mehrpooya

Affiliation: Renewable Energies and Environment Department, Faculty of New Sciences and Technologies, University of Tehran

Education: PhD Chemical engineering

Research and Professional Experience: He is head of hydrogen and fuel cell laboratory. In this laboratory they research on micro fuel cell manufacturing and nano-structured electrocatalysts. He has more than 10 years of experience in design and optimization of integrated renewable and non-renewable processes like fuel cell power plants, hydrogen production and liquefaction, high temperature solar thermal hybrid energy systems, CO_2 capturing processes and cryogenic processes. Prior to joining Renewable Energies and Environment Department he worked on various kinds of production processes especially in energy systems and catalytic processes in oil gas industries.

Professional Appointments: Ranking in the top 1% (Essential Science Indicators - Clarivate) by citations for engineering and all fields in Web of Science. Highly Cited Researcher.

Publications from the Last 3 Years:

1. Asadnia M, Mehrpooya M, Large-scale liquid hydrogen production methods and approaches: A review, *Applied Energy,* 212, 15 February 2018, Pages 57-83.
2. Fereidooni L, Tahvildari K, Mehrpooya M, Trans-esterification of waste cooking oil with methanol by electrolysis process using KOH, *Renewable Energy,* 116, 183-193, 2018.
3. Sadaghiani MS, Mehrpooya M, Ansarinasab H, Process development and exergy cost sensitivity analysis of a novel hydrogen liquefaction

process. *International Journal of Hydrogen Energy*, Available online 11 November 2017.
4. Mehrpooya M, Ansarinasab H, Sharifzadeh M. M. M, Rosen A. Process development and exergy cost sensitivity analysis of a hybrid molten carbonate fuel cell power plant and carbon dioxide capturing process, *Power sources*, 364, 2017
5. Hosseinpour S, Aghbashlo M, Tabatabaei M, Mehrpooya M, Estimation of biomass higher heating value (HHV) based on the proximate analysis by using iterative neural network-adapted partial least squares (INNPLS), *Energy* 138, 1 473-479, 2017.
6. Mehrpooya M, Sharifzade M. M. M, Conceptual and basic design of a novel integrated cogeneration power plant energy system, *Energy* 127, 15 May 2017, 516–533.
7. JalaliZonouz M, Mehrpooya M, Parametric study of a hybrid one column air separation unit (ASU) and CO2 power cycle based on advanced exergy cost analysis results, *Energy,* 140, Part 1, 2017, 261-275.
8. Mehrpooya M, Mohammadi M, Ahmadi E, Techno-economic-environmental study of hybrid power supply system: A case study in Iran, *Sustainable Energy Technologies and Assessments* 25, 1-10, 2018.
9. Mehrpooya M, Sharifzadeh MMM, Zonouz MJ, Rosen MA, Cost and economic potential analysis of a cascading power cycle with liquefied natural gas regasification, *Energy Conversion and Management,* 156, 68-83, 2018.
10. Mehrpooya M, Ashouri M, Mohammadi M, Thermoeconomic analysis and optimization of a regenerative two-stage organic Rankine cycle coupled with liquefied natural gas and solar energy, *Energy,* 126, 899–914, 2017.
11. Mehrpooya M, Dadollahi M, Modeling and investigation of high temperature phase change materials (PCM) in different storage tank configurations, *Journal of Cleaner Production*, 161, 831-839 2017.
12. Tahani M, Rabbani A, Kasaeian A, Mehrpooya M, Mirhosseini M, Design and numerical investigation of Savonius wind turbine with discharge flow directing capability, *Energy,* 130, 327–338, 2017.

13. Mehrpooya M, Ashouri Milad, Mohammadi Amin, Thermoeconomic analysis and optimization of a regenerative two-stage organic Rankine cycle coupled with liquefied natural gas and solar energy, *Energy, 126, 1 2017, 899–914.*
14. Mehrpooya M, Sharifzade M. M. M., Mahsa Rajabi, Aghbashlo M, Tabatabai M, Hosseinpour S, Ramakrishna S. Design of an integrated process for simultaneous chemical looping hydrogen production and electricity generation along with carbon capture, *International Journal of Hydrogen Energy*, In Press.
15. Hosseinpour S , Aghbashlo M , Tabatabaei M , Younesi H , Mehrpooya M , Ramakrishna S, Multi-objective exergy-based optimization of a continuous photobioreactor applied to produce hydrogen using a novel combination of soft computing techniques, *International Journal of Hydrogen Energy*, In Press.
16. Ghorbani B, Mehrpooya M, Hamedi M. H, Amidpour M, Exergoeconomic analysis of integrated natural gas liquids (NGL) and liquefied natural gas (LNG) processes, *Applied Thermal Engineering*, In Press.
17. Mehrpooya M, Esfilar R, Moosavian S. M. A, Introducing a novel air separation process based on cold energy recovery of LNG integrated with coal gasification, transcritical carbon dioxide power cycle and cryogenic CO_2 capture, *Journal of Cleaner Production*, In Press.
18. Mehrpooya M, Moftakhari Sharifzadeh M. M, A novel integration of oxy-fuel cycle, high temperature solar cycle and LNG cold recovery-energy and exergy analysis, *Applied Thermal Engineering*, In Press.
19. Mehrpooya M, Bahramin P, Pourfayaz F, Rosen M, Introducing and analysis of a hybrid molten carbonate fuel cell-supercritical carbon dioxide Brayton cycle power plant, *Sustainable Energy Technologies and Assessments* (2016), 18, 100–106.
20. Mehrpooya M, Moftakhari Sharifzadeh M. M, Modeling, energy and exergy analysis of solar chimney power plant-Tehran climate data case study, *Energy* 115 (2016) 257e273.
21. Ghorbani B, Hamedi M. H, Amidpour M, Mehrpooya M, Cascade refrigeration systems in integrated cryogenic natural gas process

(natural gas liquids (NGL), liquefied natural gas (LNG) and nitrogen rejection unit (NRU)), *Energy* 115 (2016) 88e106.

22. Mohammadi A, Mehrpooya M, Pourfayaz F, Thermodynamic and exergy analysis and optimization of a transcritical CO_2 power cycle driven by geothermal energy with liquefied natural gas as its heat sink, *Applied Thermal Engineering*. 109, Part A, 25 October 2016, Pages 640–652.
23. Mohammadi A, Mehrpooya M, Exergy analysis and optimization of an integrated micro gas turbine, compressed air energy storage and solar dish collector process, *Journal of Cleaner Production* 139, 15 December 2016, Pages 372–383
24. Mehrpooya M, Arash shafaei, Conceptual design and exergy analysis of novel flash based Helium recovery from natural gas processes, *Energy*, 114, 1 November 2016, Pages 64–83.
25. Ahmadi M. H, Mohammadi A, Pourfayaz F, Mehrpooya M, Bidi M, Valero A, Uson S, Thermodynamic analysis and optimization of a waste heat recovery system for proton exchange membrane fuel cell using transcritical carbon dioxide cycle and cold energy of liquefied natural gas, *Natural gas science and engineering*, 34, August 2016, Pages 428–438.
26. Aghaie M, Mehrpooya M, Pourfayaz F, Introducing an integrated chemical looping hydrogen production, inherent carbon capture and solid oxide fuel cell biomass fueled power plant process configuration, *Energy Conversion and Management*, 124, 15 September 2016, Pages 141–154.
27. Mehrpooya M, Lazemzade R, Sadaghiani S .M, Parishani H, Energy and advanced exergy analysis of an existing hydrocarbon recovery process, *Energy Conversion and Management* 123 (2016) 523–534.
28. Kasaeian A, Mehrpooya M, Aghaie M, Ahmadi M. H, Solar radiation prediction based on ICA and HGAPSO for Kuhin City, Iran, *Mechanics & Industry* 17 (5), 509, 2016.
29. MH Ahmadi, M Mehrpooya, N Khalilpoor, Artificial neural networks modelling of the performance parameters of the Stirling engine, *International Journal of Ambient Energy* 37 (4), 341-347.

30. Mehrpooya M, Conceptual design and energy analysis of novel integrated liquefied natural gas and fuel cell electrochemical power plant processes, *Energy,* 111, 468-483.
31. Ahmadi M. H, Mehrpooya M, Pourfayaz F, Exergoeconomic analysis and multi objective optimization of performance of a Carbon dioxide power cycle driven by geothermal energy with liquefied natural gas as its heat sink, *Energy Conversion and Management* 119 (2016) 422–434.
32. Ghorbani B, Hamedi M, Shirmohammadi Reza, Mehrpooya M, Hamedi M. H, A novel multi-hybrid model for estimating optimal viscosity correlations of iranian crude oil, *Journal of Petroleum Science and Engineering,* Vol 142, 2016, 68–76.
33. Mehrpooya M, Kalhorzahed M, Chahartaghi M, Investigation of integrated air separation process, cold energy recovery of LNG and CO_2 power cycle process configuration, *Journal of Cleaner Production,* 113, 1 2016, 411–425.
34. Mehrpooya M, Vatani Ali, Omidi M, Novel natural gas liquefaction process configuration alternatives using absorption refrigeration systems, *Applied Thermal Engineering*, 98, 5 2016, 591–604.
35. Mehrpooya M, Moftakhari Sharifzadeh M. M and Rosen M. A., Energy and exergy analyses of a novel power cycle using the cold of LNG and low-temperature solar energy, *Energy,* 95, 15 2016, 324–345.
36. Mehrpooya M, Dehghani H, Moosavian SMA. Optimal design of solid oxide fuel cell, ammonia-water single effect absorption cycle and Rankin's steam cycle hybrid system. *Journal of Power Sources*, 306, 29 2016, 107-123.

In: Liquefaction: Analysis and Assessment ISBN: 978-1-53614-773-5
Editor: Corey Hanson

Chapter 2

LIQUEFACTION HAZARD AND NUCLEAR POWER PLANT SAFETY

***Tamás János Katona**, DSc**
University of Pécs, Hungary

ABSTRACT

The severe accident at the Fukushima Dai-ichi nuclear power plant demonstrated that the secondary phenomena caused by an earthquake can cause more disastrous consequences that the vibratory ground motion itself. Moreover, even if the probability of exceedance of the external hazard effects is very low compared to the plant design basis, the nuclear power plant's (NPP) safety must be ensured by appropriate severe accident measures. Soil liquefaction can be a possible secondary phenomenon of earthquakes at soil sites of NPPs. Analysis of the liquefaction hazard and liquefaction consequences should be performed for the design basis conditions, for safety analysis of the plant and for the severe accidents. The objectives and basic requirements of these analyses are given in the chapter. The soil liquefaction hazard is a matter of acceptance of the site, therefore the hazard should be excluded from the design basis. If the potential for soil liquefaction is found to be unacceptable, the site shall be

* Corresponding Author Email: katona.tamas.janos@mik.pte.hu.

deemed unsuitable unless practicable engineering solutions are demonstrated to be available for soil stabilization. The main task is to demonstrate that plant safety can be made either via integration of the earthquake-induced liquefaction consequences into seismic probabilistic safety analysis or via adaptation of this method to the case of liquefaction. The difficulties for adaptation of the seismic probabilistic safety analysis methodology is thoroughly considered, taking into account the advances in the research of the liquefaction phenomenon and liquefaction induced building movements as well as the specific features of the structures, systems and components of the NPPs. The applicability of the state-of-the-art liquefaction risk analysis methodologies for nuclear power plants is also discussed. The possibility for assessing the nuclear power plant safety via code-deterministic margin calculation is demonstrated and could be an easy method to adapt compared to those in seismic probabilistic safety analysis. The severe accident analysis for the liquefaction case is briefly described as a logical continuation and amendment of the deterministic margin type evaluation of the safety with respect to liquefaction.

Keywords: liquefaction, nuclear power plant, safety, fragility, hazard, accident

INTRODUCTION

The catastrophe at the Fukushima-Dai-ichi Nuclear Power Plant caused by the Great Tōhoku earthquake followed by a huge tsunami on the 11th of March 2011 revealed two fundamental experiences:

The hazardous natural phenomena exceeding those accounted in the design of the nuclear power plants (NPPs) can't ever be ruled out. Consequently, the plants have to be prepared to withstand even the most disastrous effects of natural phenomena.

The plants designed in compliance with nuclear standards can survive the effects of the vibratory ground motion due to disastrous earthquake that exceeded several times the design basis one[1] [1] but may fail due to effects of phenomena accompanying or generated by the earthquakes.

---

[1] The consequences of the Miyagi earthquake (August 2005, magnitude 7.2) at the Onagawa NPP were negligible, although the recorded ground motions exceeded those the plant was designed for. July 2007 the Chūetsu offshore earthquake (Mw=6.6) hit the Kashiwazaki-Kariwa NPP.

The earthquake-induced cyclic loading of the saturated soils which may undergo either transient or permanent reduction of undrained shear resistance as a consequence of excess pore water pressures. The strength degradation may range from slight diminution of shear resistance to the seismically induced liquefaction. Liquefaction is one of those secondary effects of earthquakes that should be investigated and evaluated for soil sites of the NPPs (see e.g., [2-5]).

In this chapter the earthquake-induced liquefaction hazard, i.e., the instability of soils resulting from liquefaction and strength degradation is discussed with regard to the safety of NPPs.

As a matter of fact, the liquefaction hazard is not a generic safety issue for the NPPs. The issue is limited to the soil sites. In case of site selection for a new NPP, one of the basic questions is whether the site soil layers are susceptible to liquefaction. Generally, the soil liquefaction hazard is a matter of acceptance of the site. According to the IAEA Safety Guide SSG-35 [2], if the potential for soil liquefaction is found to be unacceptable, the site shall be deemed unsuitable unless practicable engineering solutions are demonstrated to be available. For screening out the hazard, the factor of safety to the liquefaction (FS_L) due to design basis earthquake (DBE) is applied, which is calculated by conservative methods (see e.g., [3]).

Even if the geotechnical analysis demonstrates negligible susceptibility of soils due to DBE, it does not automatically mean that the soil liquefaction could be excluded as a beyond design basis phenomenon. According to most recent requirements, accident scenarios caused either by complex failures, not accounted for in the design, or hazards with effects exceeding the design basis should be analysed. The objective of this analysis is to demonstrate that the plant safety related structures, systems and components (SSCs) have sufficient capacity to withstand the hazard effects that exceed the design

---

The experienced ground motions exceeded significantly the design basis level. After Great Tohoku earthquake, the behaviour of thirteen nuclear units in the impacted area on the East-shore of the Honshu Island demonstrated high resistance against ground vibrations due to earthquake. The stress-tests of reactors in Japan also demonstrated the capability of plants to withstand beyond design basis vibratory motion. For example, the stress-test of the Ohi NPP found that the units 3 and 4 would be able to withstand an earthquake with ground acceleration of up to 1.260 g that is exceeding 1.8 times its design basis of 0.7 g.

basis without sudden loss of function (cliff-edge effect). It has to be demonstrated that the core damage probability is very low, and the possibility of large release of radioactive substance into the environment is practically eliminated [6, 7, 8, 9]. The approach to implement these requirements vary whether operating or new plants are considered.

It has to be underlined that the safety relevance of the liquefaction hazard for NPPs is well studied, see e.g., [10] and [11]. The most recent generic summary on the phenomenon is published in [12]. Therefore, this chapter does not intend to provide new information on the liquefaction phenomenon itself. The routine procedures for dealing with the hazard in the design is also out of the scope of the chapter.

The objective is limited to the consideration of the safety issues of liquefaction as beyond design basis hazard, and the methods of how to deal with the hazard, depending on whether the safety of an operating plant or a new plant has to be assessed.

Practical motivation of the chapter is experience gained at operating plants, for example, at the Paks NPP in Hungary and some other VVER-440 plants in central and east Europe, where the site seismicity had been underestimated in the design, therefore, one of the most important tasks of the generic safety evaluation and upgrading of the plant in 1990-2000 was the re-evaluation of seismic hazards including re-evaluation of liquefaction hazard where it was appropriate, see [13, 14, 15]. At these plants, the comprehensive, complementary safety review (stress-test) demonstrated that the seismic design basis is satisfactorily determined on the basis of events consistent with a 10^{-4} per annum return frequency and the plants are robust enough to withstand the vibratory effects of earthquakes. However, the demonstration of sufficient margin to withstand the effects of soil liquefaction at some plants require essential effort [15]. These efforts had been focused on both the hazard characterisation and liquefaction failure modes of plant systems, structures and components important for safety and/or used for severe accident management, and on the effects of liquefaction on the feasibility of severe accident management.

In the chapter, both the liquefaction hazard and fragility aspects of the safety analysis for NPPs are discussed based on the literature review and studies of the author that cover the following aspects:

- interpretation of the case liquefaction hazard in the frame of defence-in-depth principle,
- applicability of the seismic probabilistic safety analysis method for the safety analysis of nuclear power plants for earthquake-induced liquefaction,
- issues of definition of fragility functions and hazard curves for probabilistic safety analysis,
- possibilities for adaptation of state-of-the-art liquefaction risk analyses for nuclear power plants,
- effects of liquefaction on the plant structures and safety functions, liquefaction failure modes of plant systems, structures and components important for safety and/or used for severe accident management,
- considerations regarding mechanisms of liquefaction induced building movements of typical structures of nuclear power plants.
- possibility for adopting the code deterministic seismic margin methodology for the case liquefaction,
- effects of liquefaction on the feasibility of severe accident management.

The issues listed above remain actual since the periodic re-evaluation of the safety of NPPs that is required for a minimum of every ten years in the majority of countries extends to the justification of the relevance of the characterization of site seismic and liquefaction hazards.

It has to be noted that the development of generic liquefaction risk evaluation methods and those specific for nuclear application are not harmonized at the moment. Significant progress has been made in liquefaction hazard assessment and risk evaluation since the publication of documents [10, 11] during last twenty years. Therefore, an attempt is made here to integrate a part of the newest achievements into the frame of safety analysis for NPPs.

Liquefaction Hazard in Context with Defence-in-Depth Principle

Design of NPPs has to follow the principles of Defence-in-Depth (DiD) [6]. The Level 1 and 2 is related to the normal operation and operational transients, therefore, these are not discussed here.

At the Level 3a that deals with cases of design basis accidents, the earthquake-induced liquefaction hazard has to be considered in relation to design basis earthquake (DBE)[2]. The DBE shall not cause liquefaction either because the soil isn't susceptible at all or the soil is improved, or the DBE excitation isn't sufficient enough for triggering the liquefaction. The evaluation of the possibility liquefaction has to be done conservatively.

At the Level 3b that is the beyond design basis condition caused by rare earthquakes[3], the liquefaction hazard should be evaluated, either because the earthquake is exceeding the DBE, or because the margin in FS_L may have turned out to be insufficient. In case of necessity the effects of liquefaction on the safety related SSCs needed to function for control of an accident such as to limit radiological releases and prevent escalation to core melt should be considered. Consequently, even if the geotechnical analysis demonstrates negligible susceptibility of soils due to DBE, it does not automatically mean that the soil liquefaction could be excluded as a beyond design basis phenomenon. This is also valid in cases when soil improvements have been implemented at the site. The design margins of SSCs considered, have to be sufficient to withstand the effects of liquefaction without loss of function. The possibility liquefaction and its consequences should be evaluated using best-estimate methods and assumptions.

---

[2] 10 CFR Part 50 Appendix S—Earthquake Engineering Criteria for NPPs (a) (ii) [2] and 10 CFR § 100.23 Geologic and seismic siting criteria (d) (4) [3].

[3] In some national regulations, see e.g. in Finnish Guide YVL B.1, it is considered as a special case of design extension condition (DEC C).

At the Level 4 of DiD, the safety objective is to control accidents with core melt to limit off-site releases using complementary safety features to mitigate core melt and implementing specific procedures for management of accidents with core melt. Liquefaction, if it happened, affect the severe accident management capabilities at the site and offsite. The most important question is whether the retaining function of the confinement would be affected by liquefaction.

At the Level 5 of DiD, the basic objective is the mitigation of consequences of significant releases of radioactive material. The question is, how the liquefaction at and around the site is hindering the off-site radiological impact necessitating protective measures.

SAFETY AND SEVERE ACCIDENT ANALYSIS ASPECTS

Safety justification of the NPPs should be carried out comprising

- design basis analysis,
- safety analysis including the analysis of beyond design basis hazard effects/consequences, and
- severe accident analysis.

The *design basis analyses* have to ensure that the design is robust in relation to the design basis earthquake with an annual probability of exceedance 10^{-4}/a $\div 10^{-5}$/a, and effective safety measures are in place. Regarding earthquake induced liquefaction and in case of new plants the hazard has to be excluded with high confidence. Compliance with this requirement can be expressed in terms of factor of safety $FS_L \gg 1$, or liquefaction potential index, $I_{LP} \ll 5$ that should be justified for DBE with 10^{-4}/a$\div 10^{-5}$/a probability of exceedance. In case of periodic or focused safety reviews of operating plants, if the liquefaction hazard can't be screened out for the DBE in light of new evidences, the procedure for beyond design basis analysis has to be utilized.

The *safety analysis* has to demonstrate that the overall risks are acceptable. The analysis of accident sequences initiated by external events with extreme low annual probability of exceedance (up-to 10^{-7}/a) is required for demonstration of low annual probability of reactor core damage ($\leq 10^{-5}$/a) or large early releases ($\leq 10^{-6}$/a). The safety analysis identifies the strengths, weaknesses and inter-dependencies in the overall design. Widely accepted methods for analyzing and demonstrating the seismic safety performance of the plant are as follows (see [16, 17] and [18]):

- the probabilistic safety analysis (PSA) that is providing the annual frequency for core damage (or early large releases of radioactive substances),
- the seismic margin analysis (SMA) that is demonstrating sufficient design margin of safety related SSCs with respect to the effects caused by earthquake.

The combination of the above methods is the "PSA-based SMA method" that models the plant response by PSA-type seismic event tree supported by fault trees. The margin of SSCs in the minimal cut set is to define in accordance with [18], i.e., via calculation of margins.

The consequences of the liquefaction should be integrated either into the seismic safety analysis (PSA, SMA), or could be analysed as a quasi-independent external event applying the methods of PSA or SMA. In the first case, there are failure modes both due to vibratory effects on the SSCs and due to the liquefaction. These failure modes can be in "either-or" or "both" relation. Under specific conditions, the SSCs are exposed to the liquefaction effects with essential time delay after the vibratory ground motion (see e.g., in [19]), therefore the liquefaction can be considered as a specific load case after the earthquake, acting on the SSCs, which just survived the effects of the earthquake vibratory motion [20]. The analysis has to evaluate the availability of these SSCs with respect to the liquefaction effects.

For the purposes of *severe accident* management, the analysis of the severe accident scenarios has to ensure provision and planning for severe but unlikely sequences, to determine those sequences for which reasonable practicable preventive or mitigatory measures can be identified. The analysis comprises a combination of engineering judgement and best estimate deterministic calculations and evaluations be made on a best estimate basis.

The basic question is whether and how the proven (in the nuclear industry) seismic safety analysis methods can be extended for the case of liquefaction.

PROBABILISTIC SAFETY ASSESSMENT FOR LIQUEFACTION HAZARD

For the assessment of the post-earthquake condition and safety of an NPP, the probability of failure of structures, systems and components (SSCs) needed for ensuring the fundamental safety functions have to be known for all failure modes. The post-event condition and damage of SSCs depends on the load vector $\boldsymbol{X}(x_1, x_2, \cdots)$ and on the conditional probability of failure, i.e., the fragility, $F(x_1, x_2, \cdots)$ with respect to these loads, and $H(x_1, x_2, \cdots)$ is the probability distribution of effects/loads.

In the probabilistic seismic safety analysis, the fragility and the hazard are defined as functions of a non-negative single load parameter, the maximum horizontal acceleration, *a* or PGA, though it is obvious that the damage of the SSCs also depends on the length of strong motion, frequency content of the vibratory motion, etc (see [21]). Following [16], the annual probability of failure of an SSC or even plant damage state, λ_{ij} can be calculated via generic equation:

$$\lambda_{ij} = \int_0^\infty \frac{dH_i(a)}{da} F_j(\mathrm{a}) da, \tag{1}$$

where $dH_i(a)/da$ is the ith representation of the annual probability density of exceedance for effects/loads, $dH_i(a)da = d\lambda(a)$ is the annual frequency for acceleration to be within the interval $[a, a + da]$, and $F_j(a)$ is the jth fragility curve, i.e., the conditional probability of failure versus PGA.

According to [16], the calculation via Equation 1 should be done for families of hazard and fragility functions, and the damage state frequency should be the weighted average value by the probabilities associated with fragility, q_i and hazard curves p_j, i.e., $\{\langle\lambda_{ij}q_i p_j\rangle\}$.

As per [16, 17], the probabilistic safety assessment of liquefaction can be integrated into the seismic PSA. Essential ingredients of the integration should follow the logic of Equation 1, i.e.,

- The hazard curve for liquefaction has to be developed, that is the function of the probability of exceedance per annum versus the engineering parameter (EDP) characterizing the safety related damaging effect of the liquefaction. In case of earthquake, that is the annual exceedance frequency versus maximum horizontal acceleration.
- The fragility function, that is the conditional probability of failure of SSCs (or even the plant as a whole) versus the safety related EDP of the hazard. Particularly, in case of earthquake, that is also the maximum horizontal acceleration.
- Plant response analysis, i.e., event tree, fault trees for modelling the plant response to the seismic effects by event tree and fault trees.

These elements of the probabilistic safety analysis for liquefaction for NPPs are within the scope of the chapter.

Issues of Fragility Evaluation for Liquefaction

Generally, the definition of fragility of SSCs consists of probabilistic characterization of the demand, D and the resistance or capacity, C, both expressed in terms of the same engineering demand parameters, $\boldsymbol{x}$ and the trivial expression for exceeding of the threshold of capacity is:

$$F(\boldsymbol{x}) = P\{\boldsymbol{D} \geq \boldsymbol{C}|\boldsymbol{x}\} \quad (2)$$

In a lucky case, a single intensity parameter, the maximum horizontal acceleration of the earthquake, *a* could be the measure for damaging potential as well, as it is in Equation 1, i.e.,

$$F(a) = \int_0^\infty P\{D \geq C|x\}\, P\{x|a\}\, dx \qquad (3)$$

If the capacity, *C,* of respective failure modes as well as the demand, *D,* are log-normally distributed with median capacities and demands (m_C and m_D respectively) and β_C and β_D are the composit logarithmic standard deviations to account for aleatoric and epistemic uncertainty of *C* and *D* respectively, the probability of failure expressed in terms of *a* will be as follows [18]:

$$P\{C - D \leq 0|a\} = \Phi\left[-\frac{ln(m_C) - ln(m_D)}{\sqrt{{\beta_C}^2 + {\beta_D}^2}}\right] \qquad (4)$$

Following the logic above, the first step for the definition of the fragility of a structure for soil liquefaction is the identification of possible failure modes of the particular structure with consequent identification of the resistance including the identification of the parameters that determine the resistance of the structure. As a matter of fact, the liquefaction affects the NPP SCCs in a complex manner. For the structures, it is a soil-foundation-structure interaction case, i.e., the resistance of the structure depends not only on the structural parameters but also of the geotechnical parameters.

The second step is the definition of the demand in terms of engineering parameters identified above and establishing the correlation between these engineering parameters with the earthquake intensity parameters, distribution of which are defined in the frame of probabilistic seismic hazard assessment. Thus, the required probabilistic characterization of the demand part of the fragility could be achieved.

For the systems and equipment inside the buildings the input set of demand parameters for the fragility definition is the response to the building structure.

There are rather limited experiences available on the safety analyses of NPPs for liquefaction hazard. In the frame of the comprehensive external hazard evaluation programme IPEEE [22] performed in the U.S., the liquefaction hazard has been considered to some extent for 41 soil sites addressing different concerns, e.g., slope instability, settlement, displacement, and stresses in buried piping. A basic finding of these analyses is that the liquefaction due to beyond design earthquake would be an essential contributor to the core damage. A similar conclusion was made on the basis of seismic PSA for Paks NPP [23].

Recently, there aren't established methodologies on how to assess and integrate the consequences into the PSA. The ASME/ANS RA-S–2008 [16] and ASME/ANS RA-Sa–2009 [17] standards are just simply stating the necessity of evaluation of consequences of liquefaction. The only and most comprehensive guidance for nuclear application is given in "Seismic Fragility Application Guide" [24]. Here, the Appendix G "General Methodology for Liquefaction Seismic Fragility Assessment and Example Analysis" shows via case study how to define the fragility curve for liquefaction. In the case study here, probabilistic fragility calculations were performed for a specific site using deterministic liquefaction potential [25] and settlement [26] evaluations and probabilistic seismic hazard analysis (PSHA). Random samples have been generated and evaluated.

The issues of the development of fragility of SSCs for liquefaction will be discussed below, recalling the findings and statements of "Seismic Fragility Application Guide" [24] which is the only guidance document for nuclear application.

In the simplest case, the occurrence of liquefaction could be the fragility and demand parameter. Analysis in [24] shows that a relatively small increment in ground-surface PGA was generally required to go from a state of incipient to gross liquefaction.

This finding would motivate the assumption that the probability of liquefaction, P_L would be more reasonably to approximate by a "no-yes" step-function, instead of the conventional double-lognormal distribution. Assuming that the failure of SSCs would depend on the occurrence of the liquefaction and, the probability distribution of failure will be the same step-function. The uncertainty could be accounted for by p-box, i.e., the family of fragility curves $\{q_i, f_i\}$, that is also applicable with Equation 1, where f_i is the ith fragility curve and q_i is the corresponding probability weight can be approximated as:

$$[\overline{P}(a), \underline{P}(a)], \tag{5}$$

where, $[\underline{P}(a) \leq P(a) \leq \overline{P}(a)]$,

$$\underline{P}(a) = \underline{P}_{fail} = \begin{cases} 0, if\ a \leq a^*|FS_L \geq 1 \\ 1, if\ a \geq a^*|FS_L \leq 1 \end{cases},$$

$$\overline{P}(a) = \overline{P}_{fail} = \begin{cases} 0, if\ a \leq a^{**}|FS_L \geq 1 \\ 1, if\ a \leq a^{**}|FS_L \leq 1 \end{cases},$$

and the a^* and a^{**} are the lowest possible and highest possible level of PGA where the gross liquefaction occurs, respectively. A similar idea was proposed e.g., in [27, 28]. However, the "no-yes" type approximation of the fragility of an SSC would be equal to a very conservative assumption of cliff-edge type damage to the SSC instantly with the incipience of liquefaction.

A little more sophisticated formula would be to use the liquefaction potential index, I_{LP} for the quantification of possible consequences of liquefaction. As it is shown for example in [29], the I_{LP} can be used for the rough and highly qualitative estimate of liquefaction consequences, e.g., $I_{LP}=5$ for modest liquefaction and $I_{LP}=15$ for severe (settlements, lateral spreads) liquefaction. In [29] real hazard curves have been developed for a liquefaction potential index that could motivate the use of I_{LP} for a damage parameter, like the PGA is used in Equation 1. However, the I_{LP} could be used just for rough and preliminary assessment of possible liquefaction effects, since the detailed correlation of damage of typical nuclear SSCs versus I_{LP} does not exist.

However, the failure of SSCs does not necessarily occur simply with the onset of liquefaction, but it is correlated with the magnitude of accumulated shear deformations and soil settlement. In [24] the level-ground soil settlement was selected for failure parameter characterizing the demand and the capacity was also expressed in terms of level of settlement. This is a practical assumption made in other cases of liquefaction analysis for NPPs, see [22]. The calculations here had been made largely based on empirical procedures.

The complexity of the soil-foundation-structure interaction problem had been investigated e.g., in [30] on the basis of centrifugal experiments. The primary mechanisms affecting the structures have been considered, e.g., localized volumetric strains due to partial drainage, consolidation due to excess pore pressure dissipation, partial bearing failure due to strength loss in the foundation soil and SSI-induced building ratcheting due to cyclic loading of foundation. The parameters affecting the mechanisms are for example, the PGA, liquefiable layer relative density and thickness, measures and parameters of the structure and the drainage of the soil. The study [30] concludes that the post-liquefaction (reconsolidation) settlement in the free-field should not be considered as an appropriate measure for the settlement, since that does not capture shear-induced deformations in the soil beneath shallow foundations. It was also found in [30] that the sufficient part of the settlement occurs during the strong shaking. The subsequent studies [31, 32, 33] identified the key parameters affecting the settlement mechanisms. In [33] the analysis has been performed using fully-coupled, 3-D, nonlinear effective stress finite element analyses for three intensity levels of shaking (measured in cumulative absolute velocity – CAV). Regarding the soil, the sensitivity study in [33] shows that the liquefiable layer thickness, depth, relative density and spatial distribution of liquefiable layers dominate the building settlement. Regarding structural parameters, the length-to-width ratio are dominating, and the foundation contact pressure and the contact area was mainly critical during stronger levels of shaking. Important findings have been achieved regarding simulation possibilities of soil-foundation-structure behaviour via fully-coupled, 3-D, nonlinear effective stress finite element analyses. The mechanisms of liquefaction-induced building movement are classified in [34] as the ground-loss due to soil ejecta, shear-induced settlement due to punching or soil-structure interaction and volumetric-induced sedimentation and consolidation settlement. A simplified procedure to estimate the building settlement and recommendations are also provided for evaluating building performance at liquefiable sites.

Although the results published in [31-34] and plenty of similar studies (e.g., [35]) are very valuable, their applicability for nuclear power plants is limited, since the structures investigated in these studies, the conditions defining the soil-foundation-structure interaction and the possible failure-modes of the structures are far different from those at the NPPs.

For example, in case of Generation II reactors, the typical diameter of a containment foundation is approximately 50 m, i.e., the contact area is an order of magnitude greater than those in the studies [30-33]. The containment building's embedment is not less than 10 m, the height-to-with ratio varies between 0.5–1.0, the foundation contact pressure is over 500 kPa far over the values in the referenced studies, and the structure is very stiff, since the containment is designed for high internal pressure and for an aircraft crash. Considering an example of an operating plant, in case of Paks NPP, Hungary, the ~10 m thick liquefiable soil layer is beneath the 150 m × 80 m large base-mat of the reactor building, the average ground-water level is about the base-mat level at ~8.5 m depth, the height to with ratio is ~0.5 and the foundation contact pressure is ~400 kPa. Except for the reactor building, all safety related buildings and connecting lifelines are founded in the upper ~10 meters thick non-liquefiable dry sand.

The experience of seismic PSA shows that the approximate number of SSCs to be accounted for in the analysis can be over 500. In case of safety analysis for liquefaction, this number should be not less: all safety related structures, especially the containment, including systems for protection of the containment, and the systems for heat removal from the core and spent fuel pool to the ultimate heat sink. An analysis of liquefaction induced behaviour of this large number and variety of structures, systems and components would cause enormous computation burden even if the simplified calculation is performed for a random sample as it has been done in [24]. Therefore, the intention for selecting single fragility parameters and as much as possible simple methods for calculation of the fragility is fully justified.

The possible failure-modes of the SSCs of NPPs are the post-liquefaction uneven settlement and tilt of certain structures resulting in loss of integrity and impeded drop of control rods into the active zone due to transient and post-liquefaction tilting, if the rods were not dropped at the early phase of strong motion (see e.g., in [20]). The tilting of building should not impede the function of systems and components to perform their safety functions.

In practical cases, the failure modes are related to transient and post-liquefaction tilting that are more dominated by the heterogeneous distribution of soil parameters and irregularities in stratigraphy rather than by the shear-type liquefaction transient tilting.

Failure of long buried or surface pipelines and channels is to be expected. These damages could be caused more by the differential between settlements of pipes and buildings at the wall penetrations [36]. Further, the communication from the emergency energy sources to their consumers and the communication to the ultimate heat sink should be practically intact.

Some earth structures (slope of dams and inlet and/or outlet channels) can also be damaged.

The most important question is whether the containment remains leak-tight which could be the case even if the containment suffers some structural damages [6, 37]. Since the plant should not be restarted after a beyond design basis earthquake plus a liquefaction event, where the limit states of SSCs are beyond the nuclear code allowances. The desired condition of the structures from the point of view of safety and accident mitigation/management are discussed e.g., in [15, 38].

From building movement mechanisms classified in [30, 34], the settlement related to the volumetric strains, particularly, the consolidation after excess pore pressure dissipation seems to be dominating in the case of the large reactor building of the Paks NPP. This can be evaluated on the basis of coupled soil-foundation-structure calculation [15].

In case of post-liquefaction consolidation settlement, the relaxation of excess pore pressure takes longer time than the duration of strong ground motion. It means the load-case liquefaction is timely separated from the load-case vibratory ground motion. This would mean that accounting for the failures due to liquefaction in seismic PSA is not fully justified.

In the case of safety related structures other than reactor building, the level-ground settlement that accounts for the inhomogeneous soil layering, i.e., enabling the definition of the possible tilting of the buildings, could be used for the fragility parameter.

Considering the simple structures, e.g., underground piping and concrete electrical ducts analyzed in [24], the calculation of median resistance with respect to the ground-level settlement is rather simple. The load on the pipe can also been expressed as a function of the ground-level settlement that can be correlated to the PGA. An important finding of [24] is that the settlement will cease at its maximum value. This reflects the physical limitation related to the limit of packing of soil particles while the PGA may monotonically increase. Consequently, the probability of failure of a structure with certain median resistance to settlement may not necessarily be equal to unity at the physical limit of soil settlement. The mathematical expression of this finding could be written as follows:

$$P(fail|a) = \int_0^{\delta_{max}} P(fail|S)P(S|a)p(a)da, \tag{6}$$

where

- $P(fail|a)$ is the conditional probability of failure for the given SSC due to liquefaction caused by the earthquake with given PGA=*a*,
- $P(S|a)$ is the conditional probability of settlement *S* versus PGA, *a*,
- $P(fail|S)$ is the conditional probability of failure versus measure of settlement, and *S* is defined as average permanent settlement of the foundation.

The probability distribution function for *S* has a cut-off at S_{max} . The $p(a)da = d\lambda_a$ is the annual probability for *a* to be in interval $[a, a + da]$. Consequently, the total probability of failure can be expressed as a function of PGA. However, even these cases are more complex than those analyzed in the guidance [24], hence the failure of the underground piping is dominated not by the settlement itself but by the differential settlement along the long pipe due to inhomogeneous soil layering and differential settlement between the pipe and structures at the wall penetrations [36].

Beside the PGA, there are earthquake intensity parameters that have been considered in plenty of studies as independent fragility variables (see e.g., [39, 40]). In [41, 42], 23 different intensity measures are listed and defined and the reasons and limitations of use of any of them are discussed depending on the objective of the investigations. In [33] CAV has been used for the earthquake intensity parameter while the effects of the soil conditions and structural features on the building settlement have been analyzed. In [39] the cumulative absolute velocity (CAV_5) was assumed to be more closely related to excess pore pressure generation in potentially liquefiable soils as the other (stress or energy related) intensity measures.

Obviously, the CAV is proportional to the product of strong motion duration and the average energy of the strong motion acceleration time history *a(t)* [43]. Thus, the CAV can be considered as the product of two random variables, the duration of strong motion *T* and the mean of absolute value of ground acceleration time history, $E\{|a(t)|\}$ [40, 43]:

$$CAV = \int_0^T |a(t)|\, dt \cong T \cdot E\{|a(t)|\} \tag{7}$$

Generally, these variables are not independent. As shown in [40] the CAV reflects the main physical properties of the cyclic load due to earthquake excitation and can be expressed as a product of duration and the average number of load-cycles in form:

$$CAV \approx \frac{1}{\pi} \cdot T \cdot A_c = 2\frac{1}{\omega_c} \cdot N \cdot A_c, \tag{8}$$

where ω_c is the median frequency of the power spectral density function of the acceleration time history, *a(t)*, A_c is the mean amplitude.

The CAV is an adequate indicator of accumulating effects of liquefaction since it is reflecting the main features of pore pressure accumulation phenomena [39] as well as the ratcheting-type settlement phenomena that are depending on the number of cycles and amplitude of cyclic motion. The higher the mean frequency of the excitation is, the less will be the possibility of a damage, which corresponds to the observations.

Use of CAV could be rather attractive, but it does not resolve the issues discussed in relation to PGA, i.e., establishing the relation between CAV and building movement requires very complex analytical procedures and needs empirical justification.

Issues Related to Definition of a Liquefaction Hazard Curve

The seismic PSA considers all sequences with very low annual frequencies up-to 10^{-7}/a. The earthquake induced liquefaction hazard should also be characterized for earthquakes with annual frequencies up-to 10^{-7}/a. For integration of the liquefaction consequences into seismic PSA, the logic of Equation 1 should be followed. The difficulties of defining the fragility as a function of a single parameter has been discussed above. Although it was concluded that the definition of the fragility of SSCs is very complex and practically impossible to define as a function of a single intensity measure of the earthquake, here we will show that the same difficulties appear regarding definition of the hazard curve.

Let's assume that the conditional probability of the average building settlement versus PGA is known and the hazard curve for PGA, $h(a) = H(a)/da$ is obtained from the PSHA, *S* has the conditional distribution function:

$$P(S) = \int_0^{\infty} P(S|a)h(a)da \quad (9)$$

It is known from [24] that $S \rightarrow S_{max}, if\ a \rightarrow \infty$. As a matter of fact, the $PGA \rightarrow a_{max}$ has also physical upper limit even due to the liquefaction. Moreover, $P(S_{max}) \leq 1$.

The state-of-the-art liquefaction hazard characterisation methods have been published in [29, 44].

The hazard in [44] is characterized by FS_L and the mean annual rate of non-exceedance of a selected factor of safety, FS_L^* has been computed as a function of earthquake intensity measures, magnitude and a_{max}. The output of the calculation is the hazard curve for each layer and the return period for $FS_L \leq FS_L^*$. However, the knowledge of this hazard curve is insufficient for the calculation of the annual failure rate of an SSC. The parameters that can be correlated with the damage as well as with the FS_L have to be identified. The considerations in the above section demonstrated the difficulties of identification of a proper damage/fragility parameter.

In [29] the liquefaction potential index has been chosen for the engineering demand parameter. The output of the calculation in [29] is the annual probability of exceedance for I_{LP}. The method is very attractive, but the application for probabilistic safety analysis of NPP is as difficult as in the case of the hazard characterisation proposed in [44].

Application of the Advanced Probabilistic Liquefaction Risk Analysis Methods for Safety Analysis of NPPs

Independent from the development of the PSA methodology for NPPs, a complex probabilistic framework has been developed for evaluation of the risk associated with earthquakes in [45, 46, 47].

The first step is the calculation of the mean annual rate for exceedance of a given level of Engineering Demand Parameters (EDP) from the conditional probability distribution P(EDP|IM) of the intensity measure (IM) that represents the parameters of the earthquake. The P(EDP|IM) also depends on the site geotechnical characteristic relevant to liquefaction. The second step is to calculate the conditional probability P(DM|EDP) for the physical effects expressed in terms of damage measures (DM) associated with the EDPs. Finally, the "risk" (decision variables – DV) associated with the DM is calculated by conditional probability P(DV|DM). The mean annual rate of exceedance of a given DV level can be calculated if the annual rate of the IM and the conditional probabilities connecting the IM to EDP, the EDP to DM and the DM to DV are known. The mean annual rate of exceedance of a given DV level, λ_{DV} can be expressed as follows:

$$\lambda_{\mathrm{DV}} = \sum_{k=1}^{N_{DM}} \sum_{j=1}^{N_{EDP}} \sum_{i=1}^{N_{IM}} P[DV|DM_k] P[DM_k|EDP_j] P[EDP_j|IM_i] \, \Delta\lambda_{IM_i}. \quad (10)$$

The triple integral form of Equation 10 is given e.g., in [48].

The procedure is applicable for the fragility analysis of SSCs and safety analysis of NPP as a whole, assuming that the parameters EDP, DM and DV and the conditional probabilities between pair of parameters can be defined.

Following the procedure in [29, 44] the magnitude and the maximum horizontal acceleration can be identified as IM, since these are the relevant measures for the definition of EDP that could be either the FS_L or I_{LP} . The difficulties appear when the damage/fragility parameter could be identified and correlated with selected EDP. The simplest choice for DM could be a measure of the settlement. However, this can't be a general damage parameter for the large number of SSCs to be considered. If it would be possible to establish the P(DM|EDP), the DV parameter could be associated to the core damage or early large release, and Equation 10 would provide the annual probability for core damage or early large release.

Unfortunately, in case of liquefaction, there are rather loose relations between the FS_L or I_{LP} as engineering demand parameters and settlement as damage parameter. Thus, the applicability of the advanced liquefaction risk analysis methods stuck on the same problems as the application of the PSA-type analysis discussed above.

Possibilities for a Margin-Type Assessment for Liquefaction

A deterministic method for seismic safety analysis of NPPs is proposed among others in the documents [16, 17, 49, 50, 51]. The methodology focuses on the seismic margin of the SSCs needed for "success," i.e., in ensuring the required safety functions. The idea is rather simple and based on the design-type rules. The elastic factors of safety can be calculated as the ratio of available capacity for sustaining the seismic load $(C - D_{NS})$ to the total seismic demand D_S calculated for a reference level earthquake, a_{RL}:

$$FS_E = \frac{C - D_{NS}}{D_S} \tag{11}$$

where *C* is the code capacity, D_{NS} is the non-seismic demand for all non-seismic loads in the load combination, and D_S the total seismic demand due to reference level earthquake with a_{RL}. The high-confidence-of-low-probability-of-failure (HCLPF) capacity is defined as:

$$HCLPF = FS_E \cdot F_\mu \cdot a_{RL}, \tag{12}$$

where F_μ accounts for the ductility. The capacity has to be defined here as Code Ultimate Strength or Code Stress Limit.

The HCLPF capacity is a measure of seismic margin defined as the earthquake motion level, at which there is a high (95%) confidence of a low (≤ 5 %) probability of failure. Using the lognormal fragility model, the HCLPF can be expressed as [16, 17]:

$$HCLPF = A_m\, exp[-1.65(\beta_R + \beta_U)], \tag{13}$$

where A_m is the median capacity, β_R and β_U are the variabilities corresponding to aleatory and epistemic uncertainties. When the composite variability βc is used, the HCLPF capacity could be approximated as the ground motion level at which the composite probability of failure is at most

1 percent. In this case, $HCLPF = A_m \, exp[-2.33\beta_C]$. This is the basis for using the HCLPF calculation in "PSA-based SMA method" [50].

The plant structures, systems and components needed for success should be identified, their HCLPF should be evaluated and the plant HCLPF should be defined via Min-Max procedure that takes the minimum HCLPF of the SSCs connected in series while ensuring the required safety functions and takes the maximum value of HCLPF of the SSCs switched parallel. The procedure is established in the documents referenced above. The case of the North-Anna NPP provided an empirical justification when the earthquake hit in August 2011 [52].

The logic could be extended to the case of liquefaction. The first step could be the selection of the reference level earthquake that could be the same as in the seismic SMA, i.e., an earthquake with maximum horizontal acceleration 0.3 g and median shaped response spectra.

Assuming that the post-liquefaction consolidation settlement is the dominating mechanism for building movement, the field of settlements has to be calculated for this reference level earthquake with subsequent evaluation of the consequences of the foundation-structure deformation. Criterion for selecting the methods for calculation of the soil-foundation-structure system is whether the soil-structure interaction has a significant impact on the liquefaction phenomenon and on the settlement. In case of soft and lightweight structures on relatively small foundations, the building follows the free-field deformations, whereas soil-structure effects must be taken into account for large heavy buildings on stiff foundations.

In the case of reactor buildings or containment of Paks NPP, a 1000 m×700 m×30 m size 3D soil model has been built. A soil-structure model has been built for static stress-field calculation. For soil-structure interaction, a 3D site response analysis has been performed. The final shear moduli and damping values have been taken from the results of the 1D analyses performed for each grid point and used as initial and final values in the 3D analyses. This calculation provided the dynamic stress field in the subsoil due to seismic shaking. For the determination of the FS_L, the static stress field and the dynamic stress field is taken into account. The FS_L is obtained by CPT-based correlation [53]. The FS_L is used for the calculation of the post-liquefaction settlements applying the method proposed in [54]. The use of different empirical methods has been analysed in several studies [55, 56], and a preference for CPT-based method has been justified. The computed settlements at the soil-foundation interface are then transformed into spring stiffness coefficients that are used for detailed calculation of the structural model.

The analysis method for other structures have been selected case-by-case. For example, in the case of plant buildings other than reactors, the structures of the free-field deformations can be used as the settlements of the building, and the free field deformations are applied as deformation loads to the foundation of the buildings. In case of underground pipelines for emergency cooling water, differential settlements between the free-field, the cooling water pump house, the turbine hall, and the Diesel generator will cause bending deformations which will be resisted by the soil surrounding the pipe. The method recommended in [57, 58] have been used for evaluation of the stress-displacement curves, which can be applied as non-linear spring foundations to analytical or numerical models of the pipe.

The brief examples above demonstrate the possibility of adaptation of the code-deterministic SMA concept for the liquefaction margin analysis, since the settlement is due to a reference level earthquake, the ultimate capacity with respect to the induced loads, as well as the demand due to building movement can be interpreted as it is required by the SMA method. The sequence of calculations is: acceleration of the reference level earthquake — FS_L — settlement field — structural loads due to foundation deformation — HCLPF. In the case of an SSC inside a safety related building, the evaluation of HCLPF will be a two-step procedure. First the structural response of the building (integrity, tilting) has to be evaluated, then the loads induced by the building deformation on the component has to be defined and the HCLPF calculation for the component has to be performed as prescribed.

The question remains open whether Equation 12 and 13 will also be valid for this case. This requires further considerations. If it would be the case, the issues discussed above with respect to the fragility and hazard calculations for liquefaction PSA could be removed.

SEVERE ACCIDENT MANAGEMENT ASPECTS

The objective of severe accident analysis is to identify severe accident scenarios, assess the accident conditions, determine those sequences for which reasonable, practicable, preventive, or mitigatory measures can be identified and designed. The analysis comprises a combination of engineering judgement and best estimate deterministic calculations and evaluations.

The task is very important for operating NPPs where the liquefaction hazard was not recognized at the phase of site investigation, since the possibility of upgrading measures are rather limited. Even if the DBE is rather moderate, the liquefaction hazard can't be automatically excluded. In spite of general belief, liquefaction can be caused by moderate earthquakes such as the earthquakes (magnitude <6.2) in Christchurch, New Zealand seen in 2010-2011. In Hungary, the M = 5.6 earthquake in 1956 at Dunaharaszti caused soil liquefaction [59]. In the case of Paks NPP located at an alluvium site, the 10^{-4}/a exceedance frequency design basis earthquake (Mw~6) could also cause liquefaction [15].

With the exception of [15, 38], the publications are very scarce on the practical severe accident analyses of NPPs for liquefaction. The analyses in [22] cover some important liquefaction caused issues. However, according to the author's knowledge, overall analysis has been made for the Paks NPP only. The procedure applied here was the same as it has been outlined for the SMA calculation above, but two important amendments have been made:

- A nonlinear soil-foundation-structure calculation has been made.
- The ultimate condition of SSCs has been defined, taking into account the required minimum safety function for severe accident management.

For the severe accident analysis, the allowable stresses and strains can be selected for the ultimate conditions similar to the LS-B in accordance with ASCE/SEI 43-05 [60] or (or LS as per FEMA 365 [61]) conditions, or in some cases, the LS-A condition in accordance with ASCE/SEI 43-05 or (or NC as per FEMA 365). These conditions can be evaluated also applying ASCE 41-13 [62] or EUROCODE 8, Part 3 [63].

It has to be emphasized, that even in the case of design extension or beyond design condition that correspond to Level 4 of the defence-in-depth, the desired condition of the containment would be to maintain the conditions as it is defined in the IAEA NS-G-1.10 Safety Guide [37], i.e., large permanent deformations and some local damages, but limited increase of leak rate that limits the deformation of the steel cladding of the containment.

CONCLUSION

A rigorous safety assessment of the nuclear power plant performed after the severe accident at Fukushima Dai-ichi NPP, demonstrated the seismic safety of NPPs and the robustness of the plants to withstand the beyond design basis ground vibratory motion. Nevertheless, the earthquakes might be the dominating contributors to the overall risk of NPPs as it has been demonstrated by the seismic probabilistic safety assessments of several NPPs. The contribution to the core damage probability of the earthquake can be strongly expressed if the associated earthquake hazard has to be accounted for, e.g., the earthquake-induced liquefaction.

Analysis of the liquefaction hazard and liquefaction consequences should be performed for the design basis conditions, for a safety analysis of the plant and for severe accidents. The objectives and basic requirements of these analyses are given in the chapter.

The main task is to demonstrate that plant safety can be made either via integration of the earthquake-induced liquefaction consequences into seismic probabilistic safety analysis or via adaptation of this method to the case of liquefaction. For both approaches the fragility of plant structures, systems and components have to be defined and a proper probabilistic characterisation of the hazard has to be done. The difficulties for adaptation of the seismic probabilistic safety analysis methodology is thoroughly considered, taking into account the advances in the research of the liquefaction phenomenon and liquefaction induced building movements as well as the specific features of the structures, systems and components of the NPPs.

The possibility for assessing the nuclear power plant safety via code-deterministic margin calculation is demonstrated that could be an easy to adapt method compared to those in seismic probabilistic safety analysis. The severe accident analysis for the case liquefaction is briefly described that is a logical continuation and amendment of the deterministic margin type evaluation of the safety with respect to liquefaction.

REFERENCES

[1] WNN, Japanese stress test results approved, World Nuclear News (Internet) 13 February 2012. *WNN*. Available from: http://www.world-nuclear-news.org/RS-Japanese_stress_test_results_approved-1302124.html. (Accessed: 2016-08-06).

[2] IAEA, Site Survey and Site Selection for Nuclear Installations, IAEA Safety Standards Series No. SSG-35. *International Atomic Energy Agency*, Vienna, 2015. ISBN 978–92–0–102415–2.

[3] Regulatory Guide 1.198 "*Procedures and Criteria for Assessing Seismic Soil Liquefaction at Nuclear Power Plant Sites*," U.S. NRC, November 2003.

[4] U.S. NRC 10CFR Part 50, *Domestic Licensing of Production and Utilization Facilities* (NRC, 1956).

[5] 10 CFR 100, Reactor Site Criteria, Appendix A "*Seismic and Geologic Siting Criteria for NPPs*," US NRC, downloaded from: http://www.nrc.gov/reading-rm/doc-collections/cfr/part100/part100-appa.html.

[6] Safety of nuclear power plants: design, SSR-2/1 Rev 1, *International Atomic Energy Agency,* Vienna, 2016. ISBN 978–92–0–109315–8.

[7] *WENRA RHWG Report WENRA Safety Reference Levels for Existing Reactors* – Update in Relation to Lessons Learned from TEPCO Fukushima Dai-Ichi Accident. (Internet) 24th September 2014 Available from: http://www.wenra.org/media/filer_public/2016/07

/19/wenra_safety_reference_level_for_existing_reactors_september_2014.pdf (Accessed: 2016-08-06).

[8] *WENRA RHWG Report – Safety of new NPP designs*, Study by Reactor Harmonization Working Group RHWG (Internet) March 2013. Available from: http://www.wenra.org/media/filer_public/2013/08/23/rhwg_safety_of_new_npp_designs.pdf (Accessed: 2016-08-06).

[9] *WENRA Statement – Safety of new NPP designs, March 2013*. Available from: http://www.wenra.org/media/filer_public/2013/04/05/wenra_statement_newdesigns2.pdf (Accessed: 2016-08-06).

[10] NUREG/CR-5741, "*Technical Bases for Regulatory Guide for Soil Liquefaction*," J. P. Koester, M. K. Sharp, M. E. Hynes, Editors, USNRC, March 2000.1.

[11] NUREG/CR-6622, "*Probabilistic Liquefaction Analysis*," USNRC, November 1999.1.

[12] National Academies of Sciences, Engineering, and Medicine. 2016. *State of the Art and Practice in the Assessment of Earthquake-Induced Soil Liquefaction and Its Consequences*. Washington, DC: The National Academies Press doi: 1017226/23474.

[13] Campbell, R. D., et al. (1998) Seismic re-evaluation of nuclear facilities worldwide: overview and status, *Nuclear Engineering and Design* 182 (1998) 17–34.

[14] Katona, T. J. Seismic Safety Analysis and Upgrading of Operating NPPs (Chapter 4), In: W*ael Ahmed (editor) Nuclear Power - Practical Aspects*, New York: InTech, 2012. pp. 77-124. ISBN: 978-953-51-0778-1.

[15] Katona, T. J. (2017). Issues of the Seismic Safety of NPPs, In: *Earthquakes – Tectonics, Hazard and Risk Mitigation*, (Ed.), InTech, DOI: 10.5772/65853. Available from: https://www.intechopen.com/books/earthquakes-tectonics-hazard-and-risk-mitigation/issues-of-the-seismic-safety-of-nuclear-power-plants.

[16] ASME/ANS RA-S-2008 "*Standard for Level 1/Large Early Release Frequency Probabilistic Risk Assessment for Nuclear Power Plant Applications*," 2008.

[17] ASME/ANS RA-Sa–2009, Addenda to ASME/ANS RA-S–2008 "*Standard for Level 1/Large Early Release Frequency Probabilistic Risk Assessment for Nuclear Power Plant Applications*," 2009.

[18] *A Methodology for Assessment of Nuclear Power Plant Seismic Margin* (Revision 1), EPRI NP-6041-SLR1, April 2013 (last amendment October 2015).

[19] Miyajima, M., Yoshida, M. and Kitaura, M. (2003), Ground Deformation and Questionnaire Survey for Withesses of Liquefaction During the 2000 Tottori-ken Siebu Earthquake, *Technical Report* MCEER-03-0003, June 30, 2003, pp. 191-201.

[20] Katona, T. J. (2015), Safety assessment of the liquefaction for NPPs, *Pollack Periodica: An International Journal for Engineering and Information Sciences* 10: (1) pp. 39-52.

[21] Katona, T. J. (2010), Options for the treatment of uncertainty in seismic probabilistic safety assessment of nuclear power plants *Pollack Periodica: An International Journal for Engineering and Information Sciences*, Volume 5: (1) pp. 121-136.

[22] NUREG-1742 Perspectives Gained from the Individual Plant Examination of External Events (IPEEE) Program, *Final Report*, Vol. 1 and 2, U.S. Nuclear Regulatory Commission, April 2002.

[23] Elter, J. (2006), Insights from the seismic probabilistic safety analysis of Paks Nuclear Power Plant. In: *Reliability, Safety and Hazard: Advances in Risk-Informed Technology,* Sanyasi Rao, V. V. S. Varde, P. V., Varde, P. V. Srividya, A., Srividya, A., & Chauhan, Ashok (Eds.). (2006). India: Narosa Publishing House Pvt Ltd.

[24] "*Seismic Fragility Application Guide*" EPRI, Palo Alto, CA: 2002. 1002988.

[25] Youd, T. L. and Idriss, I. M. (1997). "Summary Report," *Proceedings of the NCEER Workshop on Evaluation of Liquefaction Resistance of Soils* (T. L. Youd and I. M. Idriss, editors), held at Salt Lake City, Utah (January 5-6, 1996), Technical Report No. NCEER-97-0022.

[26] Ishihara, K. and Yoshimine, M., "Evaluation of Settlements in Sand Deposits Following Liquefaction During Earthquakes," *Soils and Foundations*, Vol. 32, No. 1, pp. 173-188.

[27] Katona, T. J. (2010) Options for improvement of description of nuclear power plant seismic fragility, In: ESREDA/M. Eid. *39th ESReDA Seminar on Challenges in Structural Safety & Risk Analysis*, Oct 2010, Coimbra, Portugal. pp.

[28] Katona, T. J. (2013) Options for the Treatment of Uncertainty in Seismic Safety Assessment of NPPs, *Journal of Disaster Research* 8: (3) pp. 465-472. (2013).

[29] Goda, K., Atkinson, G. M., Hunter, J. A., Crow, H. and Motazedian, D. (2011) Probabilistic Liquefaction Hazard Analysis for Four Canadian Cities, *Bulletin of the Seismological Society of America,* Vol. 101, No. 1, pp. 190–201, February 2011, doi: 10.1785/0120100094.

[30] Dashti, S., Bray, J. D., Pestana, J. M., Riemer, M., Wilson D. (2010) Mechanisms of Seismically Induced Settlement of Buildings with Shallow Foundations on Liquefiable Soil. *Journal of Geotechnical and Geoenviron. Engineering* 136:151-164.

[31] Karimi, Z. and Dashti, S. (2015b). Numerical and Centrifuge Modeling of Seismic Soil–Foundation–Structure Interaction on Liquefiable Ground, *J. Geotech. Geoenviron. Eng.*, 745 ASCE, 10.1061/(ASCE)GT.1943-5606.0001346, 04015061.

[32] Karimi, Z. and Dashti, S. (2016). Seismic Performance of Shallow Founded Structures on 747 Liquefiable Ground: Validation of Numerical Simulations using Centrifuge Experiments, *J. Geotech. Geoenviron. Eng.* ASCE, 10.1061/(ASCE)GT.1943-5606.0001479.

[33] Karimi Z., Dashti, S., Bullock Z., Porter K. and Liel A. (2018) *Key Predictors of Structure Settlement on Liquefiable Ground: A Numerical Parametric Study*, downloaded from: https://www.researchgate.net/publication/323545151.

[34] Bray, J. D. and Macedo, J. (2017) "6th Ishihara lecture: Simplified procedure for estimating liquefaction-induced building settlement," *Soil Dynamics and Earthquake Engineering*, Volume 102 – Nov 1, 2017.

[35] Karamitros, D. K., Bouckovalas, G. D. and Chaloulos Y. K. (2013) "Insight into the Seismic Liquefaction Performance of Shallow

Foundations," *J. Geotech. Geoenviron. Eng.* 2013.139: 599-607, DOI: 10.1061/(ASCE)GT.1943-5606.000079.

[36] National report of Hungary on the targeted safety re-assessment of Paks nuclear power plant, *Hungarian Atomic Energy Authority*, Budapest, 29 December 2011.

[37] Design of Reactor Containment Systems for Nuclear Power Plants, Safety Guide, *IAEA Safety Standards Series* No. NS-G-1.10, Vienna, 2004, ISBN: 92-0-103604-3.

[38] Katona, T. J., Győri, E., Bán, Z., Tóth, L. Assessment of Liquefaction Consequences for Nuclear Power Plant Paks, In: *23th Conference on Structural Mechanics in Reactor Technology*. Manchester, UK, 2015.08.10-2015.08.14. Paper ID 125.

[39] Kramer S. L., Mitchell R .A. (2006) Ground Motion Intensity Measures for Liquefaction Hazard Evaluation. *Earthquake Spectra*: May 2006, Vol. 22, No. 2, pp. 413-438.

[40] Katona, T. J., Tóth, L. Damages indicators for post-earthquake condition assessment, *Acta Geod Geophys* (2013) 48:333–345, DOI 10.1007/s40328-013-0021-9.

[41] Dashti, S., Karimi, Z. (2017), Ground Motion Intensity Measures to Evaluate I: The Liquefaction Hazard in the Vicinity of Shallow-Founded Structures, *Earthquake Spectra,* January 2017, DOI: 10.1193/103015EQS162M.

[42] Dashti, S., Karimi, Z. (2017) Ground Motion Intensity Measures to Evaluate II: The Performance of Shallow-Founded Structures on Liquefiable Ground, *Earthquake Spectra*, January 2017, DOI 10.1193/103015EQS163M.

[43] Katona, T. J. (2011), Interpretation of the physical meaning of the cumulative absolute velocity, *Pollack Periodica: An International Journal for Engineering and Information Sciences*, Volume 6, Number 1/April 2011, pp. 9-106.

[44] Steven L. Kramer, S.L. and Roy T. Mayfield, R. T. (2007) The return period of soil liquefaction, Thesis in *Journal of Geotechnical and Geoenvironmental Engineering*, July 2007, DOI: 10.1061/(ASCE) 1090-0241(2007)133:7(802).

[45] Cornell, C. A., and Krawinkler, H. (2000). "Progress and challenges in seismic performance assessment." *PEER News*, April 1–3, PEER, Berkeley, Calif.

[46] Krawinkler, H. (2002). "A general approach to seismic performance assessment." *Proc., Int. Conf. on Advances and New Challenges in Earthquake Engineering Research,* ICANCEER, Hong Kong.

[47] Deierlein, G. G., Krawinkler, H., and Cornell, C. A. (2003). "A framework for performance-based earthquake engineering." *Proc., 2003 Pacific Conf. on Earthquake Engineering*, Wellington, New Zealand.

[48] Lu, D. G., Yu, X. H. & Jia, M. M. (2012) Analytical Formulations of Fragility Functions with Applications to Probabilistic Seismic Risk Analysis, In: *15th World Conference on Earthquake Engineering 2012, (15WCEE)*, Lisbon, Portugal 24-28 September 2012Volume 1 of 38, pp. 13150-13160, ISBN: 978-1-63439-651-6.

[49] "*A Methodology for Assessment of Nuclear Power Plant Seismic Margin*" (Revision 1) EPRI NP-6041-SL, Revision 1, Project 2722-23, Final Report, August 1991.

[50] Kennedy, R. P. "Overview of Methods for Seismic PRA and Margins Methods Including Recent Innovations," *Proceedings of the Organization for the Economic Cooperation and Development/ Nuclear Energy Agency Workshop on Seismic Risk*, August 10–12, 1999, Tokyo, Japan.

[51] IAEA TECDOC-1834, "Assessment of vulnerabilities of operating nuclear power plants to extreme external events," *International Atomic Energy Agency, Vienna* 2017. ISBN 978–92–0–108817–8.

[52] Virginia Electric and Power Company (Dominion) (2011) North Anna power station units 1 and 2, North Anna independent spent fuel storage installation—summary report of August 23. *2011 earthquake response and restart readiness determination plan.* Virginia Electric and Power Company, Richmond, Virginia 23261, September 17, 2011. https://www.dom.com/about/stations/nuclear/northanna/pdf/Earthquake_Summary_Report_and_Restart_Plan_091711.pdf.

[53] Youd, T. L., at al (2001) Liquefaction Resistance of Soils: Summary Report from the 1996 NCEER and 1998 NCEER/NSF Workshops on Evaluation of Liquefaction Resistance of Soils. *J. Geotech Geoenviron. Eng.* 127(10):817–833.

[54] Zhang, G., Robertson P. K., Brachman R. W. I. (2002) Estimating liquefaction-induced ground settlements from CPT for level ground. *Can Geotech J* 39(5):1168–1180.

[55] Győri, E., Katona, T. J., Bán, Z., Tóth, L. Methods and uncertainties in liquefaction hazard assessment for NPPs, In: *Proceedings of Second European Conference on Earthquake Engineering and Seismology,* Istanbul, Turkey, 25-29.08.2014, pp. 535-546, http://www.eaee.org/Media/Default/2ECCES/2ecces_eaee/1234.pdf .

[56] Győri E, Tóth L, Gráczer Z, Katona T. (2011), Liquefaction and post-liquefaction settlement assessment — A probabilistic approach, *Acta Geodaetica et Geophysica Hungarica* 46:(3) pp. 347-369.

[57] *ASCE Technical Council on Lifeline Earthquake Engineering* (1984) Differential Ground Movement Effects on Buried Pipelines. Guidelines for the Seismic Design of Oil and Gas Pipeline Systems.

[58] American Lifelines Alliance (2001) *Guideline for the Design of Buried Steel Pipe.* http://www.americanlifelinesalliance.org/pdf/Update061305.pdf as of 17.11.2014.

[59] Győri, E., Gráczer, Z., Tóth, L., Bán, Z. T., Probabilistic liquefaction hazard analysis at liquefied sites of 1956 Dunaharaszti earthquake, in Hungary, *Geophysical Research Abstracts* 19: Paper EGU2017-7474. (2017), EGU General Assembly 2017. Vienna, Austria: 2017.04.23 - 2017.04.28.

[60] ASCE/SEI 43-05 *Seismic Design Criteria for Structures, Systems, and Components in Nuclear Facilities*, Published by American Society of Civil Engineers, 1801 Alexander Bell Drive, Reston, Virginia 20191 ISBN 0-7844-0762-2.

[61] *Prestandard and Commentary for the Seismic Rehabilitation of Buildings*, Federal Emergency Management Agency FEMA 356/November 2000, http://www.conservationtech.com/FEMA-publications/FEMA356-2000.pdf.

[62] ASCE 41-13 *Seismic Evaluation and Retrofit of Existing Buildings*, Published by American Society of Civil Engineers ISBN-13: 978-0-7844-0884-1.

[63] EN 1998-3 EUROCODE 8: *Design of structures for earthquake resistance -Part 3: Assessment and retrofitting of buildings*, 15 March 2005, https://law.resource.org/pub/eu/eurocode/en.1998.3.2005.pdf.

In: Liquefaction: Analysis and Assessment
Editor: Corey Hanson
ISBN: 978-1-53614-773-5

Chapter 3

SIMULATION AND OPTIMIZATION OF THE PROPANE PRE-COOLED MIXED REFRIGERANT (C3MR) PROCESS USING EXERGY-BASED METHOD

Eko Primabudi**[*]*, ***Liliana Calderon and Tatiana Morosuk*****[†]**
Institute for Energy Engineering, Chair of Exergy-Based Methods for Refrigeration Systems, KT Building, Berlin
Technische Universität Berlin, Germany

ABSTRACT

The uncertainties in the LNG market is currently growing, as the new producers, such as the United States and Australia emerges and the traditional importers such as Japan is shifting towards other fuels and renewables. Nonetheless, the growth of LNG trade is still proven to be healthy at a 0.5% increase over the last four years. Most of the LNG is

[*] Corresponding Author Email: e.primabudi@iet.tu-berlin.de.
[†] Corresponding Author Email: tetyana.morozyuk@tu-berlin.de.

traded with traditional long-term contract, as current importers, mostly in Asia, seeking for the security of supply. On the liquefaction side, Propane pre-cooled mixed refrigerant, known as C3MR, is the most dominant technology utilized in LNG plants. It is reported that 37% of the existing commercial plants used C3MR, whereas C3MR/SplitMR and AP-X® process, the extensions of C3MR design, accounts for 21% and 13%, respectively. In order to tackle the market uncertainties, process optimization is an essential part of the LNG project, giving the plant a competitive advantage in the long-term trade. In this case, exergy-based method is demonstrated on design assessment and optimization for both thermodynamic and cost objectives. The result is a high-resolution analysis of the thermodynamic performance of the system, in which the irreversibilities are revealed. Subsequently, the exergoeconomic analysis of C3MR process is implemented in depth by combining the economic analysis and exergy analysis. Finally, the optimization is achieved by combining the analyses with genetic algorithm technique in order to obtain the optimum cost of the liquefied natural gas. The design variables chosen for optimization are based on the exergy destruction data provided from the exergy analysis. It is shown that the optimization approach resulted in lower energy consumption by nearly 8%, exergetic efficiency improvement of 2.8% and a reduction of cost of product by 9.3%.

Keywords: LNG, C3MR, exergy, exergoeconomic, optimization

INTRODUCTION

Among one of the fossil fuel demand, natural gas has been steadily on the rise globally. The world natural gas consumption from the past 20 years revealed a high growth at 6.1%, while in the next five years it is predicted keep growing at 1.6% annually due to its abundant supply, mature technology and, more importantly, the low carbon footprint compared to other types of fossil fuels. The natural gas trade involves not only pipeline distribution but also in a liquefied form known as liquefied natural gas (LNG). In the recent years, LNG has been penetrating a sizable market share of global demand. New players such as Australia and the United States have been building new facilities to increase their LNG production, competing with the existing main suppliers such as Russia and Qatar. The new investment in LNG resulted in price uncertainties resulting in shorter purchase contracts. Nevertheless, LNG trade is steadily growing by 0.5% over the last four years since more countries are taking the LNG consumption for their industries and power generations.

At the end of 2012, aside from the seven primary liquefaction technologies, there were other technologies used sporadically through the globe [1]. The most significant share in the liquefaction market are the processes engineered by Air Products and Chemicals Inc., which include the C_3MR, AP-X, C_3MR/Split MR. The C_3MR technology has been predominantly implemented in the global LNG industry, owning the largest market share due to its versatility and capability to adapt to the process requirements in different scenarios [2]. Over 95% of the base-load LNG plants operate on mixed refrigerant processes, with the remaining few operating on conventional cascade processes. The propane pre-cooled mixed refrigerant liquefaction cycle, also known as the C_3MR cycle is one of the seven primary liquefaction technologies found worldwide today. Due to its versatility and capacity to apply to a broader range of process requirements, it has grown worldwide since its first commercial application in Brunei in 1972. C_3MR cycle is currently the most popular liquefaction technology, gaining 65% market share of the world's liquefaction capacity.

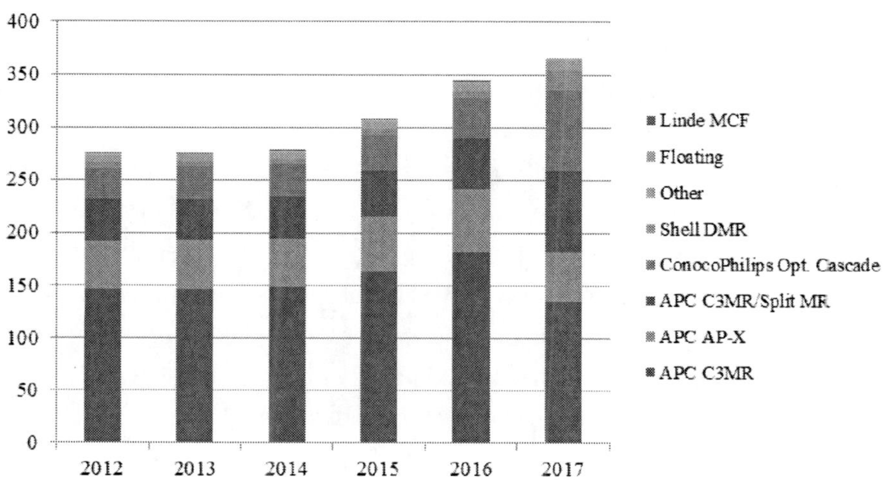

Figure 1. Liquefaction Capacity by Type of Technology, 2012-2017 [1].

The design of C3MR has also been extended to the AP-X® and SplitMR® LNG Process in order to accommodate larger liquefaction capacity. Using the AP-X®, a baseload plant capacity can be increased by more than 50%, by adding additional nitrogen refrigeration loop to the existing C3MR. This new technology has been proven in six AP-X trains currently operating in Qatar, where the nitrogen cycle subcooling loop provides refrigeration capacity up to 7.8 MTPA, the AP-X enable the additional capacity equivalent to 1-2 MTPA [3]. Meanwhile, the power split between precooling refrigeration and liquefaction refrigeration poses a challenge for fully utilizing the available power from two identical gas turbine drivers. In order to solve this issue, a portion of the liquefaction power can be shifted to the driver for the precooling refrigeration by strategically matching compressor bodies with drivers. This machinery configuration is known as SplitMR® technology. The liquefaction plants that have been installed worldwide based on their technology over the five years is depicted in Figure 1.

C3MR PROCESS

A typical process flow of the C3MR is shown in Figure 1. Propane serves as the pre-cooling medium for the natural gas feed whereas the MR is used as the primary refrigerant, configured on separated cycle. The natural gas feed from the acid gas cleaning unit is pre-cooled to about -30 to -35°C in the pre-cooling stage. The pre-cooling cycle is performed in throttling and flashing processes at different pressure levels, where it then exchanges the heat with the natural gas and MR. The propane is ultimately vaporized entirely and redirected to the propane compressors. The number of pressure levels at which propane is evaporated usually depends on the temperature of the cooling water or air. Four stages are advantageous in warm climates, while three stages are adequate in cold climates [4]. When the mixed refrigerant exits the propane cycle, it is partially condensed, and its phases have to be separated in a high-pressure separator. The vapor part is also called the light mixed refrigerant (LMR), and the liquid part is called the heavy mixed refrigerant (HMR). The cold temperature is produced with the vaporization of the mixed refrigerant with two throttling processes to subcool and liquefy the natural gas to between -155oC to -162oC by MR before store it into the LNG tanks. The MR itself exited from the main cryogenic heat exchanger (MCHE) and recompressed to complete the MR cycle. The C3MR process is versatile since it uses single refrigerant for pre-cooling, while in the MCHE, it can match the cooling curve of MCHE, which is the MR boiling curve versus the natural gas condensation curve, hence minimizing irreversibilities in the process.

Although the technology has already been adequately designed, it is essential for the LNG project owners to optimize the liquefaction plants based on their requirements and constraints. In a given processing capacity, the design and operation variables are the determining factors for the efficiency and product cost of the plant. The composition of mixed refrigerant, the pressure levels and the flow rate of refrigerants are the most important variables in this process. They will determine the appropriate component size, and ultimately, the cost of the components. The project owners not only want to have a high-efficiency plant but also to reach the least cost of product. Exergy-based method suits perfectly to perform such analysis.

Using energy analysis to evaluate a system is inadequate since the only inefficiencies detected by the energy analysis are the energy streams out of the system that are not used further. The exergy concept complements and enhances an energy analysis by calculating the true thermodynamic value of an energy carrier, known as exergy value and the real thermodynamic inefficiencies in a system, known as exergy destruction [5]. It allows engineers to evaluate the performance of the component level, both in a thermodynamic and economic sense. On each stream, it calculates the exergy values and the specific cost per exergy unit which ultimately provide the information about exergetic parameters of each component. More details about exergy-based analysis will be described in the next section.

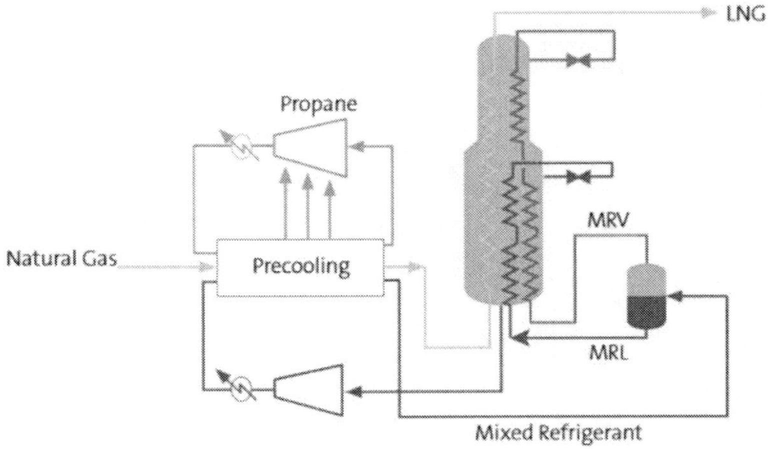

Figure 2. Air Product and Chemicals (APCI) C3MR Process [3].

LNG has been the main subject of numerous studies. However only a few focused on the optimization of the C3MR process. Moreover, there are no studies that comprehensively performed the exergy-based analysis for the C3MR process. Al-Sobhi et al. [6] investigated a multi-product system that includes LNG, GTL, and methanol production network, in which the C3MR process is implemented. The main goal is to maximize profit by simulating the upstream and downstream network simultaneously and finding the right amount of natural gas that distributed along each production line. The simulation and economic analysis were conducted with Aspen Plus and Aspen Economic Analyzer, while linear programming was used for the optimization model.

Sanavandi and Ziabasharhagh [7] have recently developed an optimization of C3MR using HYSYS optimizer functions and a method in which they attempted to obtain the closest fit of the cryogenics composite curve. The authors focus on the energy consumption, set 10 variables to the optimization, including MR composition, discharge pressure, and pre-cooling outlet temperature. The result is at the optimized condition; the overall energy consumption decreases by 5.35% from its initial condition of 1.03 MJ/kg_{LNG} to 0.97 MJ/kg_{LNG}. They concluded that the mixed refrigerant composition is the most important variables for optimizing the process.

Hatcher et al. [8] developed a slightly different approach to the LNG optimization, on which they compared eight unique objective functions and observed the most-effective one. It was found that from the design perspective, minimizing the cost of compressors should be chosen, whereas from the operational perspective, minimizing the net present value (NPV) should be favored as the objective function when there is no constraint of land usage. Similarly, Wang et al. [9] performed four different objective functions to C3MR and C3MR with split propane (C3MR-SP) process based on the APCI design [10]. The objective functions include shaft work, two different exergy efficiency expressions, and operational expenditure (OPEX). They considered the pressure and flow rate of refrigerants, the composition of mixed refrigerants (MR), and the outlet temperature of MCHE as the decision variables. The ultimate goal is to see which objective function that require the least total shaft work. According to their results, maximizing the exergy efficiency will produce the most optimum solution.

Furthermore, Alabdulkarem et al. [11] simulated the C_3MR process in Aspen HYSYS® and optimized it with the use of genetic algorithm (GA) from Matlab. This method of optimization was selected among others, due to its ability to reach a global optimum with any problem [14]. In order to reduce the complexity of the optimization task, the optimization was carried out in two stages. The two-stage optimization process included fist the optimization of the mixed refrigerant cycle and then the propane cycle. The manipulated variable was the composition of the mixed refrigerant, where it is calculated to produce energy savings from the shaft work at about 13.28% reduction in the MR cycle and 17.16% in the propane cycle compared to the base case designed by Mortazavi et al. [12]. The authors also compared the result with other similar studies [4,11] and carried out the sensitivity analysis for different pinch temperatures. It was found that the optimum process is obtained at the pinch temperature of 1K. Finally, Khan et al. [13] used non-linear programming together exergy efficiency analysis. The design variables manipulated were the refrigerant composition and flow rate, suction and evaporation pressures, and refrigerant vaporization. They achieved an improvement of efficiency, but their approach to exergetic efficiency is different from the one used in this study.

EXERGY ANALYSIS

Exergy is defined as the maximum theoretical useful work (shaft work or electrical work) obtainable as the system is brought into complete thermodynamic equilibrium with the thermodynamic environment, while the system interacts with this environment only. It consists of four parts: the physical exergy $\dot{E}^{PH}$, chemical exergy $\dot{E}^{CH}$, potential exergy $\dot{E}^{PT}$ and kinetic exergy $\dot{E}^{KN}$ [14]. The total exergy of a system j is defined as

$$\dot{E}_j = \dot{E}^{PH} + \dot{E}^{CH} + \dot{E}^{PT} + \dot{E}^{KN} \quad (1)$$

In the energy conversion system, the exergy composes mainly of two part: the physical and chemical exergy, where others, which at rest relative to the environment $e^{PT} = e^{KN} = 0$. Furthermore, one of the neglected parameters in the energy analysis is the environment pressure and temperature. The physical and chemical exergy account their effect to the maximum theoretical work of a system from one state to another. The physical exergy is defined as

$$\dot{E}_j^{PH} = \dot{m}e_j^{PH} = \dot{m}\left[\left(h_j - h_0\right) - T_0 * \left(s_j - s_0\right)\right] \quad (2)$$

$$\dot{E}_j^{PH} = \dot{m}e_j^{PH} = \dot{m}\left[\left(h_j - h_0\right) - T_0\left(s_j - s_0\right)\right] \quad (3)$$

where $\dot{m}$ is the mass flow rate, h_j and s_j is the enthalpy and entropy of system j, respectively, and T_0 is the environment temperature. Here the specific physical exergy e_j^{PH} is used as it is convenient to work with on mass or molar unit.

Physical exergy at given temperature T_i and pressure p_i can be further splitted into thermal and mechanical exergy. The thermal exergy is the maximum theoretical useful work obtainable relative to its environment temperature T_0. The mechanical exergy is the maximum theoretical useful work obtainable relative to its environment pressure p_0 [15].

$$e_j^{PH} = e_j^T + e_j^M \quad (4)$$

$$e_j^T = (h_j - h_{j\,T_0,p_i}) - T_0(s_j - s_{j\,T_0,p_i}) \quad (5)$$

$$e_j^M = (h_{j\,T_0,p_i} - h_{j\,T_0,p_0}) - T_0(s_{j\,T_0,p_i} - s_{j\,T_0,p_0}) \quad (6)$$

The specific chemical exergy $\bar{e}_l^{CH}$ an ideal mixture of N ideal gases is defined as

$$\bar{e}_j^{CH} = \sum_{l=1}^{N} x_l\, \bar{e}_l^{CH} + \bar{R}T_0 \sum_{l=1}^{N} x_l * ln(x_l) \quad (7)$$

Moreover, for the solutions of liquids (with activity coefficients γ_l) is formulated as

$$\bar{e}_j^{CH} = \sum_{l=1}^{N} x_l\, \bar{e}_l^{CH} + \bar{R}T_0 \sum_{l=1}^{N} x_l * ln(\gamma_l x_l) \quad (8)$$

where $\bar{R}$ is the ideal gas constant and x_l is the mole or mass fraction of component l in the mixture.

Furthermore, the exergy balance of a system is defined as exergy of fuel, which is the resources that required to provide the exergy of product.

$$\dot{E}_{F,sys} = \dot{E}_{P,sys} + \dot{E}_{D,sys} + \dot{E}_{L,sys} \quad (9)$$

where $\dot{E}_{P,sys}$ represents exergy of product, $\dot{E}_{D,sys}$ represents the entropy generation or irreversibilities of the system and the exergy loss, $\dot{E}_{L,sys}$, is related to the exergy transfer of dissipative components that are flow out from the given system but not used further within.

EXERGOECONOMIC ANALYSIS

A combination of exergy and economic analysis is particularly useful for evaluating thermodynamics and economic parameters simultaneously. The term exergoeconomics is first coined by Tsatsaronis [16] as the extension of exergy analysis based on the same fuel and product rule. Over the recent years, exergoeconomic analysis has been broadly implemented in numerous studies for energy conversion systems [17–20]. By using this method, the cost of investment can be assigned to each component, and the cost of each stream can be associated with its exergetic properties. For instance, the cost of exergy destruction of a component shows how exergy destruction affected not only in thermodynamics sense but also the economics of the system. The exergoeconomic balance of each component is determined using the fuel-product approach

$$c_{P,k}\dot{E}_{P,k} = c_{F,k}\dot{E}_{F,k} + \dot{Z}_k \tag{10}$$

The cost rate for the capital and operation and maintenance investment $\dot{Z}_k$ is formulated as

$$\dot{Z}_k = \frac{CC_L + OMC_L}{\tau} \times \frac{C_{BM,k}}{\sum C_{BM,k}} \tag{11}$$

where CC_L, OMC_L, τ, CBM_k are the annualized investment cost, the operation and maintenance cost, the average capacity factor, the bare module cost of kth component, respectively. In addition to the equations above, each component needs to have auxiliary costing equations to solve the system of linear equations. Further details about fuel and product rule as well as modeling the exergy and exergoeconomic balance can be found in [14, 21].

In order to minimize the total cost, it is two key parameters are useful: $\dot{Z}_k$ and the cost of exergy destruction $\dot{C}_{D,k}$. The cost rate of exergy destruction is defined as the specific cost of fuel of the kth component within a system multiplied by its exergy destruction

$$\dot{C}_{D,k} = c_{F,k}\dot{E}_{D,k} \tag{12}$$

and for the total system, it is given by

$$\dot{C}_{D,sys} = c_{F,sys}\sum \dot{E}_{D,k} \tag{13}$$

In order to evaluate the magnitude of cost contributions from those parameters, exergoeconomic factor f_k is a convenient indicator. A high value implies that the cost of capital investment of the component dominates the total cost of the component, and vice versa. It is given by

$$f_k = \frac{\dot{Z}_k}{\dot{C}_{D,k} + \dot{Z}_k} \tag{14}$$

SIMULATION AND EXERGY ANALYSIS OF C3MR

The process flowsheet of C3MR was created and simulated using Aspen Plus is illustrated in Figure 2. In the pre-cooling cycle, propane is compressed using a four-stage compression (PROP-C1 to PROP-C4) with one aftercooler while the MR is compressed in three-stage process (MR-C1 to MR-C3) with intercooling. Four heat exchangers represent precooling of propane cycle (PHX1 to PHX4), and two heat exchangers (MHX1 and MHX2) represent the MCHE. The composition of natural gas feed and base case value of MR are listed in Table 2. At the pre-cooling stage, throttling process of propane is carried out to maintain the desired pre-cooling temperature, where natural gas and MR streams exit at -33°C. Furthermore, the gaseous and liquid streams of MR are separated and pass through the throttling valves to provide the cooling effect and to liquefy the natural gas to -139°C. The last step is conducted using a throttling valve (NGTV) to further subcooled the natural gas temperature to -160°C at near-atmospheric pressure. At this stage, the LNG ready to be stored.

Table 1. Base case inputs of the C3MR simulation

Specification	Value
Natural gas mass flow rate	158.42 kg/s
Natural gas feed temperature	300 K
Natural gas feed pressure	65 bar
LNG temperature	113 K
LNG pressure	1.2 bar
Propane mass flow rate	442.7 kg/s
Mixed refrigerant compressors discharge pressure	2.5 – 5.1 – 7.2 – 14.3 bar
Propane compressors discharge pressure	7.5 – 17.5 – 48.6 bar

Table 2. Initial mass fraction of the C3MR simulation

Component	Natural gas	Mixed refrigerant
Propane	0.021	0.213
Nitrogen	0.041	0.07
Methane	0.875	0.418
Ethane	0.055	0.299
N-Butane	0.005	0.00015
I-Butane	0.003	0.00015

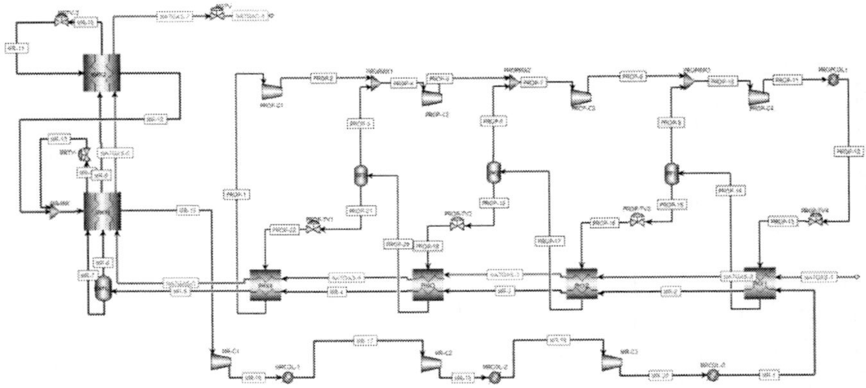

Figure 3. C3MR process modeled in Aspen Plus.

Table 1, adopted from [13], presents the initial inputs and serves as the base case for the analysis. The composition of natural gas feed and MR used in the base case simulation are listed in

Table 2. Furthermore, the simulation is modeled using the following assumption:

1) Pressure drops in the heat exchangers were assumed to be 3% (For simplification, it was assumed that there is no pressure loss for propane).
2) Isentropic efficiency was assumed to be 78% for propane compressors and 75% for mixed refrigerant compressors.
3) All compressors were assumed to have a 90% mechanical efficiency.
4) Phase separators and mixers were assumed to cause no pressure loss, while all phase separation processes were specified to simulate without heat duty.
5) The initial value for natural gas feed was adjusted according to a typical C3MR process train liquefied natural gas production capacity. In this sense, LNG production capacity is taken as 4.5 MTPA or 158.4 kg/s.
6) Base case values for the MR compositions are based on [4, 22] and were modified when necessary to make the simulation converge. The natural gas is assumed to be already pre-treated.

The thermodynamic data from the initial simulation is subsequently extracted for the exergy analysis, using the method that is explained in the previous section. The exergy balance of each component is formulated and reused as the basis for exergoeconomic analysis, where the calculation of cost per exergy unit to all streams and components is performed. It is also essential to split the exergy stream to its mechanical and thermal parts since the definition of product in liquefaction is different from energy conversion plant such as power plants. This approach ensures the accuracy of the analysis.

In this process, the exergy balance of each component is formulated with exergy of fuel and product approach [14, 21]. By referring to the process flowsheet in Figure 3, exergetic efficiency for the C3MR process is formulated with the following equation

$$\varepsilon = \frac{\dot{E}_{P,sys}}{\dot{E}_{F,sys}} = 1 - \frac{\dot{E}_{D,sys}}{\dot{E}_{F,sys}} \tag{15}$$

where,

$$\dot{E}_{P,sys} = \dot{E}^{T}_{NATGAS-7} - \dot{E}^{T}_{NATGAS-1} \tag{16}$$

$$\dot{E}_{F,sys} = \textstyle\sum_N \dot{W}_{comp,n} + \dot{E}^{M}_{NATGAS-1} - \dot{E}^{M}_{NATGAS-7} \tag{17}$$

In our case, the feed gas temperature, pressure, and the flow rate are kept constant. Consequently the exergy of product of the system $\dot{E}_{P,sys}$ is at a fixed value. Therefore, maximizing exergetic efficiency means the systems' exergy of fuel has to be reduced, which largely depends on the total energy consumption.

As the prerequisite of the exergoeconomic analysis, the exergy and economic analysis have to be initially calculated. Assumptions made for the economic analysis are the following:

1) The economic life of the plant is assumed to be 20 years.
2) All costs that contribute to the economic analysis increase with a general inflation rate of 5% per year.
3) The average annual increase in fuel cost for the power supply of the plant via compressors is 6%.
4) The average plant capacity factor τ is assumed at 85%, which means the plant operates 7446 h out of total 8760 h. available in a year. The design and construction period of the plant is assumed to be 2 years.
5) Mixers are assumed to be an integral part of the piping work; thus the purchased equipment costs are zero.

6) The average cost of power generation c_w which supplies the compressors is assumed to be 23.441 \$/GJ.

The cost of fuel and product of the system are formulated based on the process flowsheet in Figure 3:

$$\dot{C}_{F,sys} = \sum_N c_w \dot{W}_{comp,n} + c^M_{NG-1}\dot{E}^M_{NG-1} - c^M_{NG-7}\dot{E}^M_{NG-7} \tag{18}$$

$$\dot{C}_{P,sys} = c^T_{NG-7}\dot{E}^T_{NG-7} - c^T_{NG-1}\dot{E}^T_{NG-1} \tag{19}$$

The result of exergy and exergoeconomic analyses for C3MR is presented in Table 3. The exergy analysis has shown the most significant exergy destruction comes from several main components and relatively dispersed. The exergetic efficiency of 53.4% is relatively high. Nevertheless, it is revealed by the exergy analysis that the system can still be improved. The cost rate associated with investment and operation and maintenance $\dot{Z}_{sys}$ dominates the cost of product by more than 70%. This means that there is an opportunity to improve the equipment size in order to obtain a lower cost of product. Table 4 decomposes the result of analysis into the component level, where it can be seen that the biggest exergy destruction is produced by MR compressors at 30% due to high pressure required for the liquefaction in MCHE. The exergy destruction of coolers is almost 15% due to the high temperature from compression and also 15% from throttling valves due to the pressure loss to achieve low temperature condition required by the process.

The cost rate associated with exergy destruction $\dot{C}_D$ predominantly comes from MCHE at 39%, followed coolers at 20.5% and the MR compressors at 13.6% as revealed in Table 4. However, when it comes to the investment cost rate $\dot{Z}$ the pre-cooling heat exchangers are the main contributors with almost 41% from $\dot{Z}_{sys}$, followed by the MR compressors and MCHE at 37.8% and 28.7%, respectively.

Table 3. Overview of the analysis results of all cases

Specific energy consumption (MJ/t_{LNG})	0.986
Exergetic efficiency, ε (%)	53.4%
Total exergy of fuel, $\dot{E}_{F,sys}$ (MW)	239.20
Total exergy of product, $\dot{E}_{P,sys}$ (MW)	127.68
Total exergy destruction, $\dot{E}_{D,sys}$ (MW)	111.52
Total cost rate of investment and O&M, $\sum \dot{Z}_k$ ($/h)	35406
Total cost rate of fuel, $\dot{C}_{F,sys}$ ($/h)	14937
Total cost rate of product, $\dot{C}_{P,sys}$ ($/h)	50343

PROCESS OPTIMIZATION

The variables of optimization are selected according to the result obtained from the analysis. Given a fixed capacity of the feed gas, the decision variables, which consist of the MR composition, the pressure level of pre-cooling compressors and the refrigerant flow rate of can be appropriately configured to achieve optimum performance. There are many interactions between components in this process, and therefore the optimization cannot be achieved by improving a single component. Due to non-linear and non-convex character, it is also impractical to apply gradient method since this will require rigorous mathematical modeling that should match the precision from the simulation. Heuristic methods such as genetic algorithm (GA) is a popular optimization to study complex system since it does not require mathematical modeling while still can achieve a near-optimum solution. The algorithm performs a search for the optimal value by imitating the process of biological systems, where the best species is selected based on natural selection. The individuals which contain decision variables are generated iteratively based on the previous "best" individuals, also known as the elites. During the iterative loop, the elites are retained and ranked based on the value of its objective function. The mutations of the variables are also implemented to some degree in order to avoid local minima/maxima. GA programming is written in Python connected with Excel VBA and Aspen Plus. The workflow of C3MR optimization is illustrated in Figure 4, while further details regarding GA can be found in many publications such as from Holland and Goldberg [23, 24].

Table 4. Exergy Destruction, Cost Rate of Exergy Destruction and Investment of C3MR

Aspen Plus ID	Component Type	ED [MW]	$\dot{C}_D$[\$/h]	$\dot{Z}$ [\$/h]	$\dot{C}_D + Z$ [\$/h]
PHX	Pre-cooling Heat Exchangers	10.41	2637	14476	17113
MHX	MCHE	16.55	9717	7143	16861
TV	Throttling Valve	16.66	2836	258	3094
PROPC	Propane Compressors	12.65	1178	3599	4777
MRC	MR Compressors	33.53	3373	9401	12774
COL	Coolers	19.97	5095	529	5624
MIX	Mixers	1.75	46	0	46

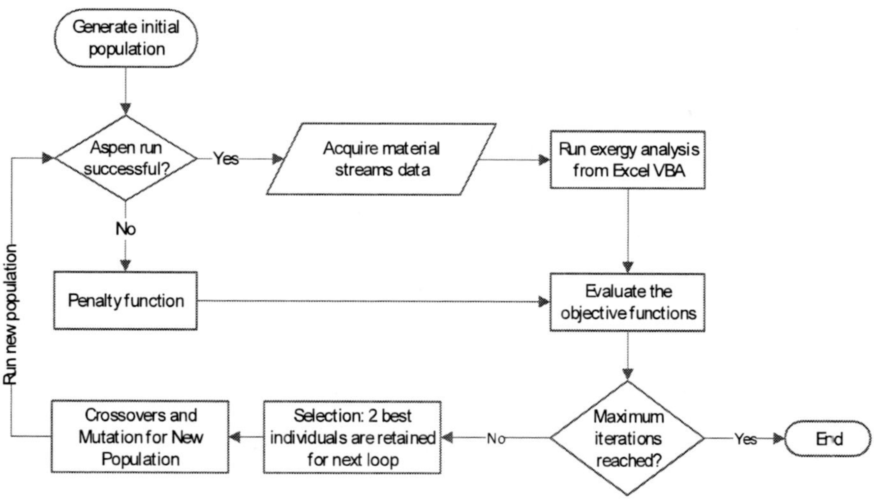

Figure 4. Workflow of the simulation and optimization of C3MR Process.

The objective function is selected by its fitness, either for minimization or maximization problem. In this case, iterations are set to 100, and the mutation rate is set to 20%. The objective is to minimize the cost of product derived from the exergoeconomic analysis. The specific energy consumption is reduced to 0.911 MJ/tLNG, which is nearly 8% reduction from the initial case. The ratio for pre-cooling is 1/3 and for MR cycle requires 2/3 of the total power consumption. From the exergetic efficiency perspective, the system has also been improved to 56.2% from previously 53.4% since it requires less exergy of fuel. The cost of product is reduced from 50343 \$/h to 45765 \$/h. The optimization has successfully optimized the two most important parameters of the system with the aid of exergy-based method.

Table 5. Overview of the analysis results of all cases

Specific energy consumption (MJ/tLNG)	0.911
Exergetic efficiency, ε (%)	56.2%
Total exergy of fuel, $\dot{E}_{F,sys}$ (MW)	227.20
Total exergy of product, $\dot{E}_{P,sys}$ (MW)	127.68
Total exergy destruction, $\dot{E}_{D,sys}$ (MW)	99.54
Total cost rate of investment and O&M, $\sum \dot{Z}_k$ (\$/h)	31745
Total cost rate of fuel, $\dot{C}_{F,sys}$ (\$/h)	13927
Total cost rate of product, $\dot{C}_{P,sys}$ (\$/h)	45672

Table 6. Exergy and Exergoeconomic Analysis Results of C3MR after Optimization

Aspen Plus ID	Component Type	ED [MW]	$\dot{C}_D$[\$/h]	$\dot{Z}$ [\$/h]	$\dot{C}_D + Z$ [\$/h]
PHX	Pre-cooling Heat Exchangers	13.58	3469	11152	14621
MHX	MCHE	10.48	5728	7618	13346
TV	Throttling Valve	17.68	3063	245	3308
PROPC	Propane Compressors	12.26	1155	3511	4667
MRC	MR Compressors	30.97	3345	8717	12062
COL	Coolers	13.96	3800	502	4301

Table 7. Decision Variables of C3MR before and after optimization

Variables	Unit	Initial Values	Optimized
MR Compressor outlet pressure	bar	48.6	48.3
MR flow rate	kg/s	301.8	285.7
MR composition			
Methane	mass fraction	0.248	0.118
Ethane	mass fraction	0.332	0.499
Propane	mass fraction	0.347	0.284
Nitrogen	mass fraction	0.072	0.099
MRTV1 pressure outlet	bar	2.9	2.9
MRTV2 pressure outlet	bar	3.0	3.0
Pre Cooling			
Propane flow rate	kg/s	442.7	444.3
Compressor 1 pressure outlet	bar	2.5	2.6
Compressor 2 pressure outlet	bar	5.1	5.4
Compressor 3 pressure outlet	bar	7.2	7.2
Compressor 4 pressure outlet	bar	14.3	11.6
PROP-TV1 pressure outlet	bar	1.3	1.4
Natural Gas			
Feed flow rate	kg/s	158.4	158.4
PHX4 outlet temperature	K	240.0	240.0
LNG pressure outlet	bar	1.2	1.2

In the component level, the cost rate associated with investment $\dot{Z}$ is still dominated by pre-cooling heat exchangers MR compressors and MCHE, albeit at a slightly lower rate. In the optimized system, these components make up for 87% of total investment cost, while on the initial case it was 88%. Due to high investment costs involved in C3MR technology, the difference between the two cases is 3533 $/h, which is actually very significant from the economic perspective. Meanwhile, the cost rate exergy destruction $\dot{C}_{D,sys}$ predominantly comes from MCHE at 28%, followed by coolers and pre-cooling heat exchangers at 18% and 17%, respectively. GA configured the variables such that the investment cost can be reduced, particularly for pre-cooling heat exchangers and MCHE. Meanwhile, the exergy destruction of MCHE, coolers and MR Compressors are also greatly reduced compared to the initial case. The decrease especially significant in the $\dot{C}_D + Z$ of MCHE, where it is almost 21% lower than the initial case. Ultimately, the total cost of product $\dot{C}_{P,sys}$ for the optimized case has been improved by 9.3%.

The decision variables before and after the optimization are presented in Table 7. Significant changes take place, as expected, in the mixed refrigerant composition and flow rate, which after optimization contains more ethane and less methane and propane. MR flow rate that passes through the cycle has been reduced by 5.3% less than the initial value. The pressure levels in the pre-cooling cycle have been modified except for the pressure outlet of compressor 3. The pressure outlet of compressor 4 (PROPC-4 in Figure 3) is altered to 11.6 bar, which produced less exergy destruction and cost of investment.

CONCLUSION

C3MR is the most popular technology in the liquefaction industry, especially for base load, mid and large size plant. The extended design of C3MR such as AP-X® and SplitMR® for 5 MTPA capacity or above have also been dominant over the years. AP-X® uses an additional nitrogen cycle to increase the processing capacity while SplitMR provides a solution for expanding the capacity by splitting the power for pre-cooling and liquefaction cycles, which share the compression drivers equally to both cycles. With over 350 MTPA LNG plant installed across the globe, C3MR is dominating at 65%.

Exergy-based method, namely exergy and exergoeconomic analysis is a convenient method to analyze complex energy system such as C3MR in order to extract the cost and exergy information at a higher resolution. In this case, exergy analysis is able to reveal the exergy destructions at the component level as well as the cost associated with each exergy stream. Furthermore, GA optimization for a complex system is a practical approach to obtain the optimum or near-optimum solution. The exergetic efficiency and cost of product prove to be the most critical parameters in evaluating the C3MR process.

Compared to the initial case, GA optimization has improved all key parameters of the system. The energy consumption is around 1/3 for pre-cooling and 2/3 for MR cycle with nearly 8% improvement. On the initial case, the exergetic efficiency and cost of product of the system is 53.4% and 50343 $/h, respectively, while after optimization the exergy analysis showed a better efficiency at 56.2% and lower cost of product at 45672 $/h. On a component level, the main contributor to investment cost are the pre-cooling heat exchangers, the MR compressors, and the MCHE. The cost associated with exergy destruction $\dot{C}_{D,sys}$ mostly dominated by MCHE and coolers. The cost of product was ultimately reduced by improving the mixed refrigerant composition and pressure level, such that the $\dot{C}_D + \dot{Z}$ of pre-cooling heat exchangers are significantly reduced from 17113 to 14621 $/h. The great improvement has also occured to $\dot{C}_D + \dot{Z}$ of MCHE from 16861 $/h to 13346 $/h.

REFERENCES

[1] International Gas Union. *World LNG Report*–2013 Edition. 2013.

[2] Pillarella M, Liu Y-N, Petrowski J, Bower R. The C3MR liquefaction cycle: versatility for a fast growing, ever-changing LNG industry. *Fifteenth Int. Conf. LNG,* Barcelona, Spain, 2007, p. 24–7.

[3] Bukowski J, Liu YN, Boccella S, Kowalski L. Innovations in natural gas liquefaction technology for future LNG plants and floating LNG facilities. *Int. Gas Union Res. Conf.*, vol. 2011, Seoul, South Korea; 2011.

[4] Venkatarathnam G, Timmerhaus KD. *Cryogenic mixed refrigerant processes*. vol. 100. Springer; 2008.

[5] Tsatsaronis G, Cziesla F. *Exergy and thermodynamic analysis* 2004.

[6] Al-Sobhi SA, Alfadala HE, El-Halwagi MM. Simulation and energy integration of a Liquefied Natural Gas (LNG) plant. *Adv Gas Process Proc 1st Int Gas Process Symp* Elsevier 2009:131–5.

[7] Sanavandi H, Ziabasharhagh M. Design and comprehensive optimization of C3MR liquefaction natural gas cycle by considering operational constraints. *J Nat Gas Sci Eng* 2016;29:176–87. doi:10.1016/j.jngse.2015.12.055.
[8] Hatcher P, Khalilpour R, Abbas A. Optimisation of LNG mixed-refrigerant processes considering operation and design objectives. *Comput Chem Eng* 2012;41:123–33. doi:10.1016/j.compchemeng.2012.03.005.
[9] Wang M, Khalilpour R, Abbas A. Operation optimization of propane pre-cooled mixed refrigerant processes. *J Nat Gas Sci Eng* 2013;15:93–105. doi:10.1016/j.jngse.2013.09.007.
[10] Roberts MJ, Ott CM. (12) Patent Application Publication (10) Pub. No.: US 2002/0187020 A1, 2010.
[11] Alabdulkarelabdulkarem A, Mortazavi A, Hwang Y, Radermacher R, Rogers P. Optimization of propane pre-cooled mixed refrigerant LNG plant. *Appl Therm Eng* 2011;31:1091–8.
[12] Mortazavi A, Somers C, Alabdulkarem A, Hwang Y, Radermacher R. Enhancement of APCI cycle efficiency with absorption chillers. *Energy* 2010. doi:10.1016/j.energy.2010.05.043.
[13] Khan MS, Lee S, Lee M. Optimization of single mixed refrigerant natural gas liquefaction plant with nonlinear programming. *Asia-Pacific J Chem Eng* 2012;7:S62–70.
[14] Bejan A, Tsatsaronis G, Moran M. *Thermal design and optimization.* John Wiley & Sons; 1996.
[15] Morosuk T, Tsatsaronis G. Graphical models for splitting physical exergy. *Shaping our Future Energy System,* vol. 1, 2005, p. 377–84.
[16] Tsatsaronis G. Exergoeconomics: Is it only a new name? *Chem Eng Technol* 1996;19:163–9. doi:10.1002/ceat.270190210.
[17] Petrakopoulou F, Tsatsaronis G, Morosuk T. Exergoeconomic Analysis of an Advanced Zero Emission Plant. *J Eng Gas Turbines Power* 2011;133:113001. doi:10.1115/1.4003641.
[18] Morosuk T, Tesch S, Hiemann A, Tsatsaronis G, Bin Omar N. Evaluation of the PRICO liquefaction process using exergy-based

methods. *J Nat Gas Sci Eng* 2015;27:23–31. doi:10.1016/j.jngse.2015.02.007.

[19] Morosuk T, Tsatsaronis G. Comparative evaluation of LNG – based cogeneration systems using advanced exergetic analysis. *Energy* 2011;36:3771–8. doi:10.1016/j.energy.2010.07.035.

[20] Tsatsaronis G. Thermoeconomic analysis and optimization of energy systems. *Prog Energy Combust Sci* 1993;19:227–57. doi:10.1016/0360-1285(93)90016-8.

[21] Lazzaretto A, Tsatsaronis G. SPECO: A systematic and general methodology for calculating efficiencies and costs in thermal systems. *Energy* 2006;31:1257–89. doi:10.1016/j.energy.2005.03.011.

[22] Gaumer L, Newton C. *Combined Cascade and Multicomponent Refrigeration System and Method.* 2447, 1973.

[23] Holland JH. Genetic algorithms. *Scientific American.* vol. 267; 1992.

[24] Goldberg DE. *Genetic algorithms.* Pearson Education India; 2006.

BIBLIOGRAPHY

A project guide to earthquakes

LCCN	2010030945
Type of material	Book
Personal name	O'Neal, Claire.
Main title	A project guide to earthquakes / Claire O'Neal.
Published/Created	Hockessin, Del.: Mitchell Lane Publishers, c2011.
Description	47 p.: col. ill., col. maps; 25 cm.
ISBN	9781584158707 (lib. bd.)
LC classification	QE521.3 .O54 2011
Contents	Earthquake! - Egg Earth -- Graham cracker faults -- Stovetop convection -- Plate tectonics: continents in motion -- Elastic earthquakes -- Seismic slinky -- Measuring the Earth's vibrations -- Monitoring the quaking Earth -- Earthquake effects: the 2004 Indian Ocean tsunami -- Earthquake effects: a better building -- Earthquake effects: liquefaction and the 1995 Kobe earthquake -- Predict the next "big one" -- Get ready for an earthquake - What to do during the earth quake.
Subjects	Earthquakes--Experiments--Juvenile literature.

	Science projects--Juvenile literature.
Notes	Includes bibliographical references (p. 44-45) and index.
Series	Earth science projects for kids

Advances in cryogenic engineering: transactions of the Cryogenic Engineering Conference, CEC: Anchorage, Alaska, USA; 17-21 June 2013

LCCN	2016448870
Type of material	Book
Meeting name	Cryogenic Engineering Conference (2013: Anchorage, Alaska)
Main title	Advances in cryogenic engineering: transactions of the Cryogenic Engineering Conference, CEC: Anchorage, Alaska, USA; 17-21 June 2013 / Chief Technical Editor: J.G. Weisend II.
Published/Produced	Melville, N.Y.: American Institute of Physics, 2014.
Description	2 volumes.
ISBN	9780735412026 (print; vol. A) 0735412022 (print; vol. A) 9780735412033 (print; vol. B) 0735412030 (print; vol. B) 9780735412019 (print; set) 0735412014 (print; set)
Variant title	Transactions of the Cryogenic Engineering Conference
Related names	Weisend, J. G., editor.
Contents	Part A. Fluid mechanics and heat transfer I; Hydrogen systems; Instrumentation I; JT coolers; Large scale systems, facilities and testing I; Superfluid helium; Large scale refrigeration and liquefaction I; Pulse tube and stirling cryocoolers (aerospace) I; LTS magnet systems I; Thermal

insulation systems; Pulse tube and stirling cryocoolers (aerospace) II; Regenerators; Fluid mechanics and heat transfer II; European spallation source; Large scale refrigeration and liquefaction II; Multilayer insulation systems; HTS magnet systems; Thermoacoustic refrigerators and drives; SRF systems I; ITER; Distribution systems -- Part B. Instrumentation II; Large scale refrigeration and liquefaction III; Pulse tubes; Cryogenic power cables, leads and fault current limiters I; Magnetic, sorption, and other coolers; Fluid mechanics and heat transfer III; Industrial and medical applications; GM cryocoolers; Large scale systems, facilities and testing II; Expanders, pumps, and compressors I; Large scale refrigeration and liquefaction IV; Aerospace applications; SRF systems II; Liquefaction, storage, and use of cryofuels; Below 4 K cryocoolers; Stirling type pulse tube cryocoolers; Expanders, pumps, and compressors II; Cryogenic systems; Fluid mechanics and heat transfer IV; LTS magnet systems II; Cryogenic safety and reliability; Modeling I; Stirling and GM cryocoolers; Large scale refrigeration and liquefaction IV; Cryogenic power cables and leads II; Instrumentation III; Modeling II; Tactical cryocoolers for space use; SRF systems III.

Subjects	Low temperature engineering--Congresses.
Notes	Includes bibliographical references and indexes.
Series	Advances in cryogenic engineering; 59A-59B
	AIP conference proceedings, 0094-243X; volume 1573
	Advances in cryogenic engineering; 59A-59B.
	AIP conference proceedings (Online); no. 1573.

Alternative fuels: current status, proposals for new standards, and related infrastructure issues: hearing before the Subcommittee on Energy and Air Quality of the Committee on Energy and Commerce, House of Representatives, One Hundred Tenth Congress, first session, May 8, 2007.

LCCN	2009376523
Type of material	Book
Corporate name	United States. Congress. House. Committee on Energy and Commerce. Subcommittee on Energy and Air Quality.
Main title	Alternative fuels: current status, proposals for new standards, and related infrastructure issues: hearing before the Subcommittee on Energy and Air Quality of the Committee on Energy and Commerce, House of Representatives, One Hundred Tenth Congress, first session, May 8, 2007.
Published/Created	Washington: U.S. G.P.O.: For sale by the Supt. of Docs., U.S. G.P.O., 2008.
Description	v, 190 p.: ill.; 24 cm.
ISBN	9780160826269 0160826268
LC classification	KF27 .E5524 2007k
Subjects	Alcohol as fuel--United States. Biodiesel fuels--United States. Coal liquefaction--United States. Renewable energy sources--United States. Alternative fuel vehicles--United States.
Notes	Shipping list no. unavailable. "Serial no. 110-42."

Alternative transportation fuels: an overview: hearing before the Subcommittee on Energy and Air Quality of the Committee on Energy and Commerce, House of Representatives, One Hundred Tenth Congress, first session April 18, 2007.

LCCN	2009376123
Type of material	Book
Corporate name	United States. Congress. House. Committee on Energy and Commerce. Subcommittee on Energy and Air Quality.
Main title	Alternative transportation fuels: an overview: hearing before the Subcommittee on Energy and Air Quality of the Committee on Energy and Commerce, House of Representatives, One Hundred Tenth Congress, first session April 18, 2007.
Published/Created	Washington: U.S. G.P.O.: For sale by the Supt. of Docs., U.S. G.P.O., 2008.
Description	v, 119 p.: ill.; 24 cm.
ISBN	9780160823879
	0160823870
LC classification	KF27 .E5524 2007i
Subjects	Alcohol as fuel--United States.
	Biodiesel fuels--United States.
	Coal liquefaction--United States.
	Renewable energy sources--United States.
	Alternative fuel vehicles--United States.
Notes	Shipping list no. unavailable.
	"Serial No. 110-31."
	Includes bibliographical references.

Animals in emergencies: learning from the Christchurch earthquakes

LCCN	2017392537
Type of material	Book
Personal name	Potts, Annie, 1965- author.

Main title	Animals in emergencies: learning from the Christchurch earthquakes / Annie Potts, Donelle Gadenne.
Published/Produced	Christchurch, New Zealand: Canterbury University Press, 2014. ©2014
Description	xii, 274 pages: colour illustrations; 24 cm
ISBN	9781927145500 (pbk.)
LC classification	HV4890.A3 P68 2014
Related names	Gadenne, Donelle, author.
Summary	After the 7.1 magnitude earthquake that shook Canterbury on 4 September 2010, the news media were quick to report, with understandable relief, that no lives had been lost. In fact, this first quake killed at least 3000 chickens, eight cows, one dog, a lemur and 150 aquarium fish, and that was only the first of a series of even more catastrophic quakes that were to follow, in which many humans and animals perished. Animals in Emergencies: Learning from the Christchurch Earthquakes provides a record of what happened to the animals during and after these quakes, and asks what we can learn from these events and our response to them. The accounts of professionals and volunteers involved in the rescue, shelter and advocacy of the city's animals post-quakes are presented in the first part of the book, and are followed by the tales of individual animals. These accounts provide an honest and compelling historical record of how Christchurch's seismic activity affected human-animal relationships in both positive and negative ways. We share our lives with a variety of companion animals, including dogs, cats, horses, fish, birds, rats, rabbits, guinea pigs and turtles, and

	the stories of how the Canterbury earthquakes affected these animals are absorbing, sometimes heart-breaking and often heart-warming. The book also reports on the fate of urban wildlife such as hedgehogs, eels and seabirds, in the aftermath of liquefaction and other damage caused by the more than 20,000 aftershocks since the first major earthquake, and considers the particular risks to animals most vulnerable when disasters strike - those confined on farms and in laboratories.
Contents	Rescue, shelter and advocacy -- Animal earthquake stories.
Subjects	Animal rescue--New Zealand--Canterbury. Animal welfare--New Zealand--Canterbury. Pets--Social aspects--New Zealand--Canterbury. Canterbury Earthquake, N.Z., 2010. Christchurch Earthquake, N.Z., 2011.
Notes	Includes bibliographical references (pages 255-263) and index.

Applied science. Engineering & mathematics

LCCN	2015430336
Type of material	Book
Main title	Applied science. Engineering & mathematics / editor, Donald R. Franceschetti, Ph.D., the University of Memphis.
Edition	[1st ed.].
Published/Produced	Ipswich, Mass.: Salem Press, a division of EBSCO Information Services; Amenia, N.Y.: Grey House Publishing, 2013. ©2013
Description	xxi, 756 pages: illustrations; 27 cm
ISBN	9781619252448 1619252449

LC classification	TA145 .A67 2013
Variant title	Engineering and mathematics
Portion of title	Engineering & mathematics
Related names	Franceschetti, Donald R., 1947- editor.
Contents	Aeronautics and Aviation; Algebra; Antiballistic Missile Defense Systems; Applied Mathematics; Applied Physics; Architecture and Architectural Engineering; Artificial Intelligence; Artificial Organs; Audio Engineering; Avionics and Aircraft Instrumentation; Biomechanical Engineering; Bionics and Biomedical Engineering; Bioprocess Engineering; Bridge Design and Barodynamics; Calculus; Chaotic Systems; Charge-Coupled Devices; Chemical Engineering; Civil Engineering; Climate Engineering; Climate Modeling; Climatology; Coal Gasification; Coal Liquefaction; Communication; Communications Satellite Technology; Computer Engineering; Computer Graphics; Computer Languages, Compilers, and Tools; Computer Networks; Computer Science; Cryptology and Cryptography; Digital Logic; Earthquake Engineering; Earthquake Prediction; Ecological Engineering; Electrical Engineering; Electrical Measurement; Electric Automobile Technology; Electrometallurgy; Electronic Materials Production; Electronics and Electronic Engineering; Energy-Efficient Building; Energy Storage Technologies; Engineering; Engineering Mathematics; Engineering Seismology; Environmental Biotechnology; Environmental Engineering; Enzyme Engineering; Fiber-Optic Communications; Fluid Dynamics; Fossil Fuel Power Plants; Fractal Geometry; Game Theory;

Geoinformatics; Geometry; Grammatology; Hydraulic Engineering; Industrial Pollution Control; Information Technology; Integrated-Circuit Design; Internet and Web Engineering; Landscape Architecture and Engineering; Mechanical Engineering; Metallurgy; Naval Architecture and Marine Engineering; Numerical Analysis; Paper Engineering; Parallel Computing; Pattern Recognition; Petroleum Extraction and Processing; Plane Surveying; Plastics Engineering; Pneumatics; Polymer Science; Probability and Statistics; Proteomics and Protein Engineering; Pyrotechnics; Quality Control; Radio; Risk Analysis and Management; Robotics; Sanitary Engineering; Scanning Probe Microscopy; Sewage Engineering; Software Engineering; Spacecraft Engineering; Space Environments for Humans; Sports Engineering; Structural Composites; Submarine Engineering; Thermal Pollution Control; Time Measurement; Topology; Transportation Engineering; Trigonometry; Typography; Urban Planning and Engineering; Video Game Design and Programming; Virtual Reality.

Subjects Engineering.
Mathematics.
Engineering.
Mathematics.

Notes Comprises articles on engineering and mathematics extracted from the five-volume reference set Applied Science. Salem Press, 2012.
Includes bibliographical references and index.

Assessment and mitigation of liquefaction hazards to bridge approach embankments in Oregon

LCCN	2003431923
Type of material	Book
Main title	Assessment and mitigation of liquefaction hazards to bridge approach embankments in Oregon / by Stephen E. Dickenson ... [et al.]
Published/Created	Salem, OR: Oregon Dept. of Transportation, Research Group; [Springfield, Va.: Available through the National Technical Information Service, 2002]
Description	2 v.: ill., charts; 28 cm.
Links	http://www.odot.state.or.us/tddresearch/reports.htm
LC classification	TE216 .A87 2002
Related names	Dickenson, S. E. Oregon. Department of Transportation. Research Group. Oregon State University. Department of Civil, Construction, and Environmental Engineering.
Subjects	Bridge approaches--Oregon--Maintenance and repair. Soil liquefaction--Oregon. Earthquake engineering--Oregon.
Notes	"November 2002." Includes bibliographical references (p. 193-210). Final report. 1994-2001. Performed by Oregon State University Dept. of Civil, Construction, and Environmental Engineering for Oregon Dept. of Transportation, Research Group and the Federal Highway Administration under contract no. SPR 361.

Biofuels from agricultural wastes and byproducts

LCCN	2010011388
Type of material	Book
Main title	Biofuels from agricultural wastes and byproducts / edited by Hans P. Blaschek, Thaddeus C. Ezeji, Jürgen Scheffran.
Published/Created	Ames, Iowa: Wiley-Blackwell, 2010.
Description	xii, 262 p.: ill.; 25 cm.
ISBN	9780813802527 (hardback: alk. paper) 0813802520 (hardback: alk. paper)
LC classification	TP360 .B5745 2010
Related names	Blaschek, Hans P. Ezeji, Thaddeus C. Scheffran, Jürgen.
Contents	Biofuels from agricultural wastes and byproducts: an introduction / Hans P. Blaschek, Thaddeus C. Ezeji, and Jürgen Scheffran -- Production of liquid biofuels from biomass: emerging technologies / Nasib Qureshi, Stephen Hughes, and Thaddeus C. Ezeji -- Butanol production from lignocellulosic biomass / Thaddeus C. Ezeji and Hans P. Blaschek -- Practical aspects of methane production from agricultural wastes / Largus T. Angenent and Norman R. Scott -- Conversion of agricultural residues to bioethanol: the roles of cellulases and cellulosomes / Edward A. Bayer ... [et al.] -- Fuel and chemical production from glycerol, a biodiesel waste product / Syed Shams Yazdani, Anu Jose Mattam, and Ramon Gonzalez -- Farm-gate to plant-gate delivery of lignocellulosic feedstocks from plant biomass for biofuel production / Klein E. Ileleji, Shahab Sokhansanj, and John S. Cundiff -- Conversion of existing dry-mill ethanol operations to biorefineries / Timothy C. Lindsey --

	Cellulosic ethanol from agricultural residues / Bin Yang, Yanpin Lu, and Charles E. Wyman -- Hydrothermal liquefaction process to convert biomass into crude oil / Yuanhui Zhang -- Detoxification of lignocellulosic hydrolysates / Bin Wang and Hao Feng.
Subjects	Agricultural wastes as fuel.
	Biomass energy.
Notes	Includes bibliographical references and index.

Biomass conversion: the interface of biotechnology, chemistry and materials science

LCCN	2012933321
Type of material	Book
Main title	Biomass conversion: the interface of biotechnology, chemistry and materials science / Chinnappan Baskar, Shikha Baskar, Ranjit S. Dhillon, editors.
Published/Created	Berlin; New York: Springer, c2012.
Description	xxvii, 465 p.: ill.; 24 cm.
ISBN	9783642284175 (hbk)
	3642284175 (hbk)
LC classification	TP248.B55 B552 2012
Related names	Baskar, Chinnappan.
	Baskar, Shikha.
	Dhillon, Ranjit S.
Contents	Biomass Conversion to Energy / Maneesha Pande and Ashok N. Bhaskarwar -- Biomass Energy / A. K. Kurchania -- Lignocellulose Pretreatment by Ionic Liquids: A Promising Start Point for Bio-energy Production / Haibo Xie, Wujun Liu and Zongbao K. Zhao -- Application of Ionic Liquids in the Conversion of Native Lignocellulosic Biomass to Biofuels / Marcel Lucas, Gregory L.

Wagner and Kirk D. Rector -- Catalysts in Thermochemical Biomass Conversion / Ali Sınağ -- Fatty Acids-Derived Fuels from Biomass via Catalytic Deoxygenation / Bartosz Rozmysłowicz, Päivi Mäki-Arvela and Dmitry Yu. Murzin – Biobutanol: The Future Biofuel / Manish Kumar and Kalyan Gayen -- Molecular Genetic Strategies for Enhancing Plant Biomass for Cellulosic Ethanol Production / Rengasamy Ramamoorthy and Prakash P. Kumar -- Production of Bioethanol from Food Industry Waste: Microbiology, Biochemistry and Technology / V. K. Joshi, Abhishek Walia and Neerja S. Rana -- Enhancement of Biohydrogen Production by Two-Stage Systems: Dark and Photofermentation / Tugba Keskin and Patrick C. Hallenbeck -- Organosolv Fractionation of Lignocelluloses for Fuels, Chemicals and Materials: A Biorefinery Processing Perspective / Ming-Fei Li, Shao-Ni Sun, Feng Xu and Run-Cang Sun -- Lignin as Source of Fine Chemicals: Vanillin and Syringaldehyde / Paula C. Rodrigues Pinto, Eduardo A. Borges da Silva and Alírio E. Rodrigues -- Liquefaction of Softwoods and Hardwoods in Supercritical Methanol: A Novel Approach to Bio-Oil Production / J. Andres Soria and Armando G. McDonald -- Bioextraction: The Interface of Biotechnology and Green Chemistry / Rakesh Kumar Sharma, Alok Adholeya, Aditi Puri and Manab Das.

Subjects Biomass conversion.

Biomass conversion.

Notes Includes bibliographical references and index.

Cameron Liquefaction Project: draft environmental impact statement.

LCCN	2014472335
Type of material	Book
Main title	Cameron Liquefaction Project: draft environmental impact statement.
Published/Produced	Washington, DC: Federal Energy Regulatory Commission, Office of Energy Projects, [2013]
Description	1 volume (various pagings): illustrations, color maps; 28 cm
LC classification	TP350 .C24 2014
Related names	United States. Federal Energy Regulatory Commission. Office of Energy Projects, issuing body.
Subjects	Cameron LNG, LLC. Environmental impact statements--Louisiana. Liquefied natural gas pipelines--Environmental aspects--Louisiana. Liquefied natural gas shipping terminals--Environmental aspects--Louisiana--Calcasieu Parish. Liquefied natural gas shipping terminals--Environmental aspects--Louisiana--Cameron Parish.
Notes	Cover title. "FERC/EIS-248D." "DOE/EIS-0488." "Docket no. CP13-25-000; docket no. CP13-27-000." Format not distributed to depository libraries. Includes bibliographical references.
Additional formats	Online version: Cameron Liquefaction Project: draft environmental impact statement (OCoLC)869298234

Coal gasification: opportunities and challenges: hearings before the Committee on Energy and Natural Resources, United States Senate, One Hundred Tenth Congress, first session, to address opportunities and challenges associated with coal gasification, including coal-to-liquids and industrial gasification, May 24, 2007.

LCCN	2007473180
Type of material	Book
Corporate name	United States. Congress. Senate. Committee on Energy and Natural Resources.
Main title	Coal gasification: opportunities and challenges: hearings before the Committee on Energy and Natural Resources, United States Senate, One Hundred Tenth Congress, first session, to address opportunities and chalenges associated with coal gasification, including coal-to-liquids and industrial gasification, May 24, 2007.
Published/Created	Washington: U.S. G.P.O.: For sale by the Supt. of Docs., U.S. G.P.O., 2007.
Description	iii, 102 p.; 24 cm.
Links	PDF version: http://frwebgate.access.gpo.gov/cgi-bin/getdoc.cgi?dbname=110_senate_hearings&docid=f:37273.pdf
LC classification	KF26 .E55 2007x
Subjects	Coal gasification--United States. Coal liquefaction--United States. Energy development--United States. Energy policy--United States.
Notes	Distributed to some depository libraries in microfiche. Shipping list no. unavailable. Includes bibliographical references.
Additional formats	Also available via Internet from the GPO Access web site. Address as of 9/26/07: http://frwebgate.

	access.gpo.gov/cgi-bin/getdoc.cgi?dbname=110_senate_hearings&docid=f:37273.pdf
Series	S. hrg.; 110-120 United States. Congress. Senate. S. hrg.; 110-120.

Computer simulation of earthquake effects: proceedings of sessions of Geo-Denver 2000

LCCN	00042141
Type of material	Book
Meeting name	Geo-Denver 2000 (2000: Denver, Colo.)
Main title	Computer simulation of earthquake effects: proceedings of sessions of Geo-Denver 2000 / sponsored by The Geo-Institute of the American Society of Civil Engineers; edited by Kandiah Arulanandan, A. Anandarajah, Xiang Song Li: August 5-8, 2000, Denver, Colorado.
Published/Created	Reston, VA: American Society of Civil Engineers, 2000.
Description	xi, 203 p.: ill.; 22 cm.
ISBN	0784405239
LC classification	TA711.A1 G46 2000
Related names	Arulanandan, Kandiah. Anandarajah, A. Li, Xiang Song. American Society of Civil Engineers. Geo-Institute.
Subjects	Soil dynamics--Computer simulation--Congresses. Earthquakes--Computer simulation--Congresses. Soil liquefaction--Computer simulation--Congresses. Soil structure interaction--Computer simulation--Congresses.
Notes	Includes bibliographical references and indexes.
Series	Geotechnical special publication; no. 110

Constraints on JP-900 jet fuel production concepts

LCCN	2007013242
Type of material	Book
Personal name	Bartis, James T., 1945-
Main title	Constraints on JP-900 jet fuel production concepts / James T. Bartis, G. Torrance Flint, Jr.
Published/Created	Sant Monica, CA: RAND Project Air Force, 2007.
Description	viii, 13 p.: ill.; 28 cm.
Links	Table of contents only http://www.loc.gov/catdir/toc/ecip0714/2007013242.html
ISBN	9780833041449 (pbk.: alk. paper) 0833041444 (pbk.: alk. paper)
LC classification	TP343 .B327 2007
Related names	Flint, G. Torrance.
Subjects	United States. Air Force. Coal liquefaction. Jet planes--Fuel. Petroleum coke.
Notes	Includes bibliographical references (p. 13).
Series	Technical report; TR-465-AF Technical report (Rand Corporation); TR-465-AF.

Cyclic behaviour of soils and liquefaction phenomena: proceedings of the International Conference on Cyclic Behaviour of Soils and Liquefaction Phenomena, 31 March-02 April 2004, Bochum, Germany

LCCN	2004445155
Type of material	Book
Meeting name	International Conference on Cyclic Behaviour of Soils and Liquefaction Phenomena (2004: Bochum, Germany)
Main title	Cyclic behaviour of soils and liquefaction phenomena: proceedings of the International Conference on Cyclic Behaviour of Soils and

	Liquefaction Phenomena, 31 March-02 April 2004, Bochum, Germany / edited by Th. Triantafyllidis.
Published/Created	Leiden: New York A.A. Balkema Publishers, c2004.
Description	xiii, 690 p.: ill.; 26 cm.
Links	Publisher description http://www.loc.gov/catdir/enhancements/fy0653/2004445155-d.html
ISBN	9058096203
LC classification	TA711.5 .I575 2004
Related names	Triantafyllidis, Theodoros.
Subjects	Soil-structure interaction--Congresses. Soil liquefaction--Congresses.
Notes	Includes bibliographical references and index.

Design of pile foundations in liquefiable soils

LCCN	2010455099
Type of material	Book
Personal name	Madabhushi, Gopal.
Main title	Design of pile foundations in liquefiable soils / Gopal Madabhushi, Jonathan Knappett, Stuart Haigh.
Published/Created	London: Imperial College Press, c2010.
Description	xiv, 217 p.: ill. (some col.); 24 cm.
ISBN	9781848163621 1848163622
LC classification	TA780 .M33 2010
Related names	Knappett, Jonathan. Haigh, Stuart.
Subjects	Piling (Civil engineering) Soil-structure interaction. Soil liquefaction. Earthquake engineering.

Notes	Includes bibliographical references (p. 203-210) and index.

Diamino amino acids

LCCN	2008026975
Type of material	Book
Main title	Diamino amino acids / Hanna A. Schäfer and Lilli M. Wohlbier, editors.
Published/Created	New York: Nova Biomedical Books, c2008.
Description	xii, 256 p.: ill. (some col.); 27 cm.
Links	Table of contents only http://www.loc.gov/catdir/toc/fy0906/2008026975.html
ISBN	9781604568929 (hardcover) 1604568925 (hardcover)
LC classification	QP562.D53 D53 2008
Related names	Schäfer, Hanna A. Wohlbier, Lilli M.
Contents	Asymmetric dimethyl L-arginine (ADMA) and delayed cerebral vasospasm after aneurysmal subarachnoid hemorrhage (aSAH) / Carla S. Jung and Ryszard M. Pluta -- Neurotropic and immunotropic effects of amino acid l-lysine / Liudmila Severyanova and Maxim Dolgintsev -- D-serine and schizophrenia / Kenji Hashimoto -- Analysis of methylated arginines by high-performance liquid chromatography-florescence detection: a review / Makoto Tsunoda -- Yield of compatible solute hydroxyectoine through conversion from ectoine by recombinant escherichia coli / Shinichi Nagata ... [et al.] -- Significant yield of ectoine by halomonas sp. EG6 during osmotic downshock treatment / Omneya Osman ... [et al.] -- Increase in amino acids and ammonia during liquefaction of semen / Thomas

	W. Stief -- Selected diamino acids in soils of differently managed mountain meadow and forest ecosystems: assessment of their role in ecosystem nutrition / Pavel Formanek ... [et al.] -- Arginine conserves the hemostasis activation state of plasma even against freezing/thawing / Thomas W. Stief.
Subjects	Diamino amino acids.
	Amino Acids, Diamino.
Notes	Includes bibliographical references and index.

Direct thermochemical liquefaction for energy applications

LCCN	2018304120
Type of material	Book
Main title	Direct thermochemical liquefaction for energy applications / edited by Lasse Rosendahl.
Published/Produced	Duxford, United Kingdom: Elsevier: Woodhead Publishing, [2018]
Description	xiii, 356 pages; 24 cm.
ISBN	008101029X
	9780081010297
LC classification	TP360 .D57 2018
Related names	Rosendahl, Lasse, 1967- editor.
Summary	Direct Thermochemical Liquefaction for Energy Applications presents the state-of-the-art of the value chains associated with these biomass conversion technologies. It covers multiple feedstock availability and feedstock composition impact on process chemistry and product quality and composition. Expert authors from around the world explore co-processing benefits, process parameters, implementation and scaling, upgrading to drop-in liquid biofuels or integration into existing petrochemical refinery infrastructure. Finally, these topics are put into a sustainability

	perspective by establishing an LCA framework for this type of process. Its focus on implementation based on the most comprehensive knowledge makes this book particularly useful for researchers and graduate students from all sorts of background working in the field of biomass and biofuels. It is also a valuable reference for engineers working to commercialize DTL technologies, engineering specialists designing process equipment, refinery professionals and developers.
Subjects	Biomass. Biomass energy. Pyrolysis. Biomass. Biomass energy. Pyrolysis.
Notes	Includes bibliographical references and index.
Additional formats	Electronic version: Direct thermochemical liquefaction for energy applications. Duxford, United Kingdom: Elsevier Ltd.: Woodhead Publishing, [2018] 9780081010259 (OCoLC) 1012399792
Series	Woodhead Publishing series in energy Woodhead Publishing in energy.

Discussion drafts concerning energy efficiency, smart electricity grid, Energy Policy Act of 2005 Title XVII loan guarantees, and standby loans for coal-to-liquids projects: hearing before the Subcommittee on Energy and Air Quality of the Committee on Energy and Commerce, House of Representatives, One Hundred Tenth Congress, first session, May 24, 2007.

LCCN	2009376310

Type of material	Book
Corporate name	United States. Congress. House. Committee on Energy and Commerce. Subcommittee on Energy and Air Quality.
Main title	Discussion drafts concerning energy efficiency, smart electricity grid, Energy Policy Act of 2005 Title XVII loan guarantees, and standby loans for coal-to-liquids projects: hearing before the Subcommittee on Energy and Air Quality of the Committee on Energy and Commerce, House of Representatives, One Hundred Tenth Congress, first session, May 24, 2007.
Published/Created	Washington: U.S. G.P.O.: For sale by the Supt. of Docs., U.S. G.P.O., 2008.
Description	v, 114 p.: ill.; 24 cm.
ISBN	9780160825507 0160825504
LC classification	KF27 .E5524 2007j
Subjects	Energy consumption--Law and legislation--United States. Energy development--Law and legislation--United States. Power resources--Law and legislation--United States. Electric power distribution--United States. Electric power production--United States. Coal liquefaction--United States--Finance.
Notes	Shipping list no.: 2009-0203-P. "Serial no. 110-50." Includes bibliographical references.
Reproduction no./Source	552-070-38579-7 GPO

Additional formats	United States. Congress. House. Committee on Energy and Commerce. Subcommittee on Energy and Air Quality. Discussion drafts concerning energy efficiency, smart electricity grid, Energy Policy Act of 2005 Title XVII loan guarantees, and standby loans for coal-to-liquids projects v, 114 p. (OCoLC)312678247

Earthquake geotechnical engineering design

LCCN	2014931392
Type of material	Book
Main title	Earthquake geotechnical engineering design / Michele Maugeri, Claudio Soccodato, editors.
Published/Produced	Cham [Switzerland]: Springer, [2014] ©2014
Description	xii, 388 pages: illustrations (some color); 24 cm.
ISBN	3319031813 (hbk.) 9783319031811 (hbk.)
LC classification	TA654.6 E3785 2014
Summary	Pseudo-static analysis is still the most-used method to assess the stability of geotechnical systems that are exposed to earthquake forces. However, this method does not provide any information about the deformations and permanent displacements induced by seismic activity. Moreover, it is questionable to use this approach when geotechnical systems are affected by frequent and rare seismic events. Incidentally, the peak ground acceleration has increased from 0.2-0.3 g in the seventies to the current value of 0.6-0.8 g. Therefore, a shift from the pseudo-static approach to performance-based analysis is needed. Over the past five years considerable progress has been made in Earthquake Geotechnical Engineering

Design (EGED). The most recent advances are presented in this book in 6 parts. The evaluation of the site amplification is covered in Part I of the book. In Part II the evaluation of the soil foundation stability against natural slope failure and liquefaction is treated. In the following 3 Parts of the book the EGED for different geotechnical systems is presented as follows: the design of levees and dams including natural slopes in Part III; the design of foundations and soil structure interaction analysis in Part IV; underground structures in Part V. Finally in Part VI, new topics like the design of reinforced earth retaining walls and landfills are covered.

Contents Part I. Site Characterisation and site amplification -- Spatially constrained inversion of surface wave data to build shear wave velocity models / Sebastiano Foti and Laura Valentina Soco -- Site classification and spectral amplification for seismic code provisions / Anastasios Anastasiadis and Evi Riga -- Part II. Liquefaction -- Sand liquefaction observed during recent earthquake and basic laboratory studies on aging effect / Takaji Kokusho, Yohta Nagao, Fumiki Ito, and Takashi Fukuyama -- Liquefaction in Tokyo Bay and Kanto regions in the 2011 great East Japan earthquake / Kenji Ishihara, Kazuhiro Araki, and Kamata Toshiyuki -- Allowable settlement and inclination of houses defined after the 2011 Tohoku-Pacific Ocean earthquake in Japan / Susumu Yasuda -- Part III. River levees and dams -- Seismic performances of river levees; experience and prediction / Ikuo Towhata -- Earthquake performance design of dams using destructive

	potential factor / G. Rodolfo Saragoni -- Part IV. Foundations and soil-structure interaction -- Seismic responses of shallow footings: a promising application for the macro-element approach / Claudio di Prisco and Michele Maugeri -- Seismic response of a large-scale highway interchange system / Kyungtae Kim, Ahmed Elgamal, George Petropoulos, Aysegul Askan, Jacobo Bielak, and Gregory L. Fenves -- Seismic displacement based design of structures: relevance of soil structure interaction / Gian Michele Calvi, Manuela Cecconi, and Roberto Paolucci -- Part V. Underground structures -- Performance and seismic design of underground structures / Kyriazis Pitilakis and Grigorios Tsinidis -- Part VI. Special topics -- Reinforced soil walls during the recent great earthquakes in Japan and geo-risk-based design / Yoshihisa Miyata -- Performance based seismic design of geosynthetic barriers for waste containment systems / Edward Kavazanjian Jr., Mohamed Arab, and Neven Matasovic.
Subjects	Earthquake engineering--Congresses. Geotechnical engineering--Congresses.
Notes	Includes bibliographical references and index.
Series	Geotechnical, geological and earthquake engineering, 1573-6059; 28 Geotechnical, geological, and earthquake engineering; 28. 1573-6059

Earthquake hazard assessment: India and adjacent regions

LCCN	2018014709
Type of material	Book
Personal name	Kolathayar, Sreevalsa, author.

Main title	Earthquake hazard assessment: India and adjacent regions / Sreevalsa Kolathayar & T.G. Sitharam (Department of Civil Engineering, Indian Institute of Science, Bangalore, India).
Published/Produced	London, UK: CRC Press/Balkema, an imprint of the Taylor & Francis Group, [2018]
ISBN	9781315143811 (ebook)
LC classification	QE537.2.I4
Related names	Sitharam, T. G., 1961- author.
Contents	Introduction -- Earthquake data and seismic source models -- Deterministic seismic hazard assessment -- Seismicity analysis and characterization of source zones -- Probabilistic seismic hazard assessment -- Site response and liquefaction analyses.
Subjects	Earthquake hazard analysis--India.
Notes	Includes bibliographical references and index.
Additional formats	Print version: Kolathayar, Sreevalsa, author. Earthquake hazard assessment London, UK: CRC Press/Balkema, an imprint of the Taylor & Francis Group, [2018] 9781138309234 (DLC) 2018011859

Earthquake hazard assessment: India and adjacent regions

LCCN	2018011859
Type of material	Book
Personal name	Kolathayar, Sreevalsa, author.
Main title	Earthquake hazard assessment: India and adjacent regions / Sreevalsa Kolathayar & T.G. Sitharam, Department of Civil Engineering, Indian Institute of Science, Bangalore, India.
Published/Produced	Boca Raton: CRC Press/Taylor & Francis Group, [2018]
Description	xii, 176 pages: illustrations, maps; 26 cm

ISBN	9781138309234 (hbk: alk. paper)
LC classification	QE537.2.I4 K65 2018
Related names	Sitharam, T. G., 1961- author.
Contents	Introduction -- Earthquake data and seismic source models -- Deterministic seismic hazard assessment -- Seismicity analysis and characterization of source zones -- Probabilistic seismic hazard assessment -- Site response and liquefaction analyses.
Subjects	Earthquake hazard analysis--India.
Notes	Includes bibliographical references and index.
Additional formats	Online version: Kolathayar, Sreevalsa, author. Earthquake hazard assessment London, UK: CRC Press/Balkema, an imprint of the Taylor & Francis Group, [2018] 9781315143811 (DLC) 2018014709

Earthquakes

LCCN	2007004657
Type of material	Book
Personal name	Park, Louise, 1961-
Main title	Earthquakes / Louise Park.
Published/Created	North Mankato, Minn.: Smart Apple Media, c2007.
Description	32 p.: ill. (some col.), col. maps; 29 cm.
Links	Table of contents only http://www.loc.gov/catdir/toc/ecip0710/2007004657.html
ISBN	9781599201115 1599201119
LC classification	QE521.3 .P37 2007
Contents	Natural disasters -- Earthquakes -- Disaster file: Pakistan -- Earth's layers -- Tectonic plates -- Disaster file: San Andreas Fault -- During an earthquake -- Measuring an earthquake -- Disaster

	file: Chile -- Liquefaction and avalanches -- Disaster file: Peru -- Floods and tsunamis -- Disaster file: Indian Ocean -- Disaster relief -- Disaster files at a glance.
Subjects	Earthquakes--Juvenile literature.
	Natural disasters--Juvenile literature.
Notes	Includes index.
	"A+."

Energy studies

LCCN	2014412944
Type of material	Book
Personal name	Shepherd, W. (William), 1928- author.
Main title	Energy studies / W. Shepherd, Formerly University of Bradford, UK and Ohio University, USA, D.W. Shepherd, University of Bradford, UK.
Edition	Third edition.
Published/Produced	London: Imperial College Press, [2014]
	©2014
Description	xxviii, 526 pages: illustrations (some colour); 26 cm
ISBN	9781848168503 (hardback)
	1848168500 (hardback)
LC classification	TJ828 .S54 2014
Related names	Shepherd, D. W., author.
Summary	This book considers the various options of renewable energy, including water energy, wind energy and biomass, solar thermal and solar photovoltaic energy. And should the nuclear option remain open?
Contents	Machine generated contents note: ch. 1 Energy and Power -- 1.1. Energy Conversion -- 1.2. Mechanical Energy -- 1.2.1. Linear motion -- 1.2.2. Rotational motion -- 1.3. Electrical Energy -- 1.4.

Chemical Energy -- 1.5. Nuclear Energy -- 1.6 Thermal Energy -- 1.7. Thermodynamics and Heat Energy -- 1.7.1. Quantity of heat -- 1.7.2. Mechanical equivalent of heat -- 1.7.3. The first law of thermodynamics -- 1.7.4. The second law of thermodynamics -- 1.7.4.1. Ideal heat engine -- 1.7.4.2. Practical heat engine -- 1.7.4.3. Ideal reverse heat engine (heat pump) -- 1.7.5. Worked examples on thermodynamics and heat energy -- 1.8. Entropy -- 1.8.1. Entropy in heat-work systems -- 1.8.2. Entropy on a cosmic scale -- 1.9. Power -- 1.10. Units and Conversion Factors -- 1.11. Problems on Energy and Power -- References -- ch. 2 Energy Resources and Energy Use -- 2.1. Energy Input to the Earth -- 2.1.1. Solar radiation and annual variation -- 2.1.2. Terrestrial energy from inside the earth. 2.1.3. Tidal (gravitational) input energy -- 2.2. Energy Flow upon Earth from Natural Sources -- 2.3. Energy Outflow from the Earth -- 2.4. Energy Stored Within the Fossil Fuels -- 2.5. Energy Production and Consumption -- 2.5.1. Energy consumption in the world -- 2.5.2. Energy production and use in the UK -- 2.5.3. Energy production and use in the USA -- 2.5.4. World fossil fuel production and consumption -- 2.6. Risks Associated with Energy Systems -- 2.6.1. Industrial accidents and industrial diseases -- 2.6.2. Large-scale accidents and sabotage -- 2.6.3. Management of energy waste -- 2.6.4. Ecosystem effects -- 2.6.5. Water supply problems -- 2.6.6. Emissions -- 2.6.6.1. Carbon dioxide emissions -- 2.6.6.2. Gaseous emissions and the greenhouse"r; effect -- 2.6.6.3. Carbon trading -- 2.7. Summary: Where Do We Go from Here? -- 2.7.1. An energy strategy -- 2.8. Problems and Review Questions -- References -- ch. 3 Electricity -- 3.1. Introduction. 3.2. Some Basic Electrical Relationships -- 3.2.1. Voltage, current and power -- 3.2.2. Worked examples on electrical circuits -- 3.3. The Generation of Electricity -- 3.4. The Siting of Electrical Power Plants -- 3.4.1. Fuel supply -- 3.4.2. Water supply -- 3.4.3. Land elevation -- 3.4.4. Road and rail access -- 3.4.5. Height of the structures -- 3.4.6. Disposal of waste products -- 3.4.7. Proximity to populated areas -- 3.4.8. Environmental implications -- 3.5. World Electricity Consumption -- 3.6. UK Electricity -- 3.6.1.Organization in 2010 -- 3.6.2. Consumption and supply -- 3.6.3.Comparative costs of electrical generation in the UK -- 3.7. US Electricity Consumption and Production -- 3.8.Combined Heat and Power (CHP) -- 3.8.1. Technologies used in current UK CHP schemes -- 3.8.2. CHP in the UK -- 3.8.3. CHP in the USA -- 3.9. Efficient Utilization of Electrical Energy -- 3.9.1. Avoiding waste -- 3.9.2. Monitoring and control -- 3.9.3. Redesigning to reduce energy costs. 3.9.4. Maintenance of equipment -- 3.9.5. Power factor correction -- 3.9.6. Maintenance of supply current waveform -- 3.9.7. Choice and use of electric motors -- 3.9.8. Load factor -- 3.9.9. Choice of lighting systems -- 3.10. Problems and Review Questions -- References -- ch. 4 Coal -- 4.1. The History and Status of Coal -- 4.1.1. Fossil fuel resources -- 4.1.2.Composition and ranking of coal -- 4.1.3. Coal mining -- 4.2. World Reserves, Production and Consumption of Coal -- 4.2.1. World coal reserves -- 4.2.2. World coal production -- 4.2.3. World coal consumption -- 4.2.4. UK coal production and consumption -- 4.2.5. US coal production and consumption -- 4.3. Coal Transportation -- 4.3.1. Surface transportation -- 4.3.2. Coal slurry pipelines -- 4.4. Emissions and Effluents from Coal -- 4.4.1. Open coal fires -- 4.4.2. Effluents due to coal burning -- 4.4.2.1. Sulphur oxides (SOx) -- 4.4.2.2. Nitrogen oxides (NOx) -- 4.4.2.3. Particulates -- 4.4.2.4. Carbon dioxide. 4.4.2.5. Carbon dioxide emissions due to coal -- 4.4.2.6. An American case study. The side effects of a coal plant -- 4.5. Carbon Capture and Storage (CCS) -- 4.6. Advanced Coal Technologies -- 4.6.1. Fluidized-bed combustion -- 4.6.2.Combined-cycle generation -- 4.7. Liquid Fuels from Coal -- 4.7.1. Indirect liquefaction -- 4.7.2. Pyrolysis -- 4.7.3. Solvent extraction -- 4.7.4. Direct hydrogenation (catalytic liquefaction) -- 4.8. Problems and Review Questions -- References -- ch. 5 Oil -- 5.1. Geological Formation -- 5.2. Origin and Extraction of Oil -- 5.3. History and Development of the Petroleum Industry -- 5.3.1. The Seven Sisters -- 5.3.2. European oilfields -- 5.3.3. OPEC -- 5.3.4. Major oil areas -- 5.4. World Oil Reserves -- 5.5. World Production and Consumption of Crude Oil -- 5.5.1. World oil production -- 5.5.2. World oil consumption -- 5.5.3. UK oil production and consumption -- 5.5.4. US oil production and consumption -- 5.6. Unconventional Liquid Fossil Fuels. 5.6.1. Oil shale -- 5.6.1.1. Oil shale mining -- 5.6.1.2. In situ combustion -- 5.6.2. Tar sands (oil sands) -- 5.7. Environmental Issues -- 5.8. Problems and Review Questions -- References -- ch. 6 Natural Gas -- 6.1.

Nature and Origin of Natural Gas -- 6.2. History and Development -- 6.3. Natural Gas Reserves -- 6.4. Production and Consumption of Natural Gas -- 6.4.1. World natural gas production -- 6.4.2. World natural gas consumption -- 6.4.3. UK natural gas production and consumption -- 6.4.4. US natural gas production and consumption -- 6.5. Unconventional Sources of Natural Gas -- 6.5.1. Coal-bed methane -- 6.5.1.1. World reserves of coal-bed methane -- 6.5.1.2. US reserves of coal-bed methane -- 6.5.2. Natural gas hydrates -- 6.5.3. Shale gas -- 6.5.4. Tight gas -- 6.6. Environmental Aspects of Natural Gas -- 6.7. Synthetic Gas from Coal -- 6.8. Problems and Review Questions -- References -- ch. 7 Geothermal Energy -- 7.1. Physical Basis of Geothermal Energy. 7.2. Geological Structure of the Earth -- 7.3. Origin of Geothermal Heat Flow -- 7.4. Geothermal Energy Resources -- 7.5. Geothermal Reservoirs -- 7.6. Locations and Types of Principal Geothermal Sources -- 7.6.1. Dry steam sources -- 7.6.2. Wet steam sources -- 7.6.3. Hot brine sources -- 7.6.4. Dry rock sources -- 7.6.5. Molten magma -- 7.6.6. Worldwide applications -- 7.7. Geothermal Prospects in the UK -- 7.7.1. Shallow drilling -- 7.7.2. Worked example -- 7.8. Geothermal Uses in the USA and Elsewhere -- 7.8.1. Hot springs and bathing spas (balneology) -- 7.8.2. Agriculture -- 7.8.3. Aquaculture -- 7.8.4. Industry -- 7.9. Geothermal District Heating -- 7.10. Geothermal Heat Pumps -- 7.10.1. Heating -- 7.10.2. Cooling -- 7.11. Electricity Generation from Geothermal Sources -- 7.11.1. Worldwide geothermal electrical power production -- 7.11.2. Technologies of geothermal electrical power generation -- 7.11.3. Locations of geothermal electricity-generating stations. 7.12. Environmental Features of Geothermal Power -- 7.12.1. Geothermal site exploration and development -- 7.12.2. Protection of the local atmosphere -- 7.12.3. Protection of groundwater -- 7.12.4. Enhancement of reservoir water -- 7.12.5. Ecological effects of geothermal plants -- 7.12.6. Effects on local geological structure -- 7.13. Problems and Review Questions -- References -- ch. 8 Nuclear Energy -- 8.1. Basic Atomic Theory -- 8.2. Basic Nuclear Theory -- 8.2.1. Nuclear fission -- 8.2.2. Worked examples -- 8.3. Radioactivity -- 8.3.1. Nature of radioactivity -- 8.3.2. Energy and decay rate -- 8.3.3. Worked examples -- 8.4. Nuclear Radiation -- 8.4.1. Forms of radiation -- 8.4.2. Units of measurement of radiation -- 8.4.3. Effects of nuclear radiation -- 8.4.4. Sources and amounts of nuclear radiation -- 8.4.4.1. Natural radiation sources -- 8.4.4.2. Man-made sources -- 8.4.5. Uses of nuclear radiation -- 8.4.5.1. Geological dating -- 8.4.5.2. Archaeological dating. 8.4.5.3. Medical tracer elements -- 8.4.5.4. Small nuclear power packs -- 8.4.5.5. Biological effects on human tissue -- 8.5. Nuclear Reactors -- 8.5.1. Thermal (fission) reactors -- 8.5.2. Uranium supplies -- 8.5.3. Plutonium -- 8.5.4. Fast breeder reactors -- 8.5.5. Reactor safety -- 8.5.6. Nuclear reactor accidents -- 8.5.6.1. Three Mile Island -- 8.5.6.2. Chernobyl -- 8.5.6.3. Fukushima Daiichi -- 8.5.6.3.1. Consequences -- 8.5.6.3.2. Radioactive releases -- 8.5.6.3.3. Severity of the incident -- 8.5.6.3.4. Public perception -- 8.6. Nuclear Waste -- 8.6.1.

Sources of waste -- 8.6.2. Waste disposal -- 8.6.3. Terrorist action -- 8.7. Nuclear-Powered Electricity Generation -- 8.7.1. Nuclear generation in the USA -- 8.7.2. Nuclear generation in the UK -- 8.8. Nuclear Fusion -- 8.8.1. Basic theory -- 8.8.2. Nuclear fusion materials -- 8.8.2.1. Supply of deuterium -- 8.8.2.2. Supply of lithium -- 8.8.2.3. Supply of tritium -- 8.8.3. Nuclear fusion reactors. 8.8.3.1. Nuclear plasma properties -- 8.8.3.2. Terrestrial fusion -- 8.8.4. Fusion reactor research -- 8.8.4.1. JET -- 8.8.4.2. ITER -- 8.8.5. Advantages of nuclear fusion as an energy option -- 8.8.5.1. Fuel supply advantages -- 8.8.5.2. Zero emissions/green energy -- 8.8.5.3. Zero high-level nuclear waste -- 8.8.5.4. Zero risk of nuclear catastrophe -- 8.8.5.5. Modest land use and plentiful site options -- 8.8.5.6. No intermittency of supply -- 8.9. Problems and Review Questions -- References -- ch. 9 Water Energy -- 9.1. Hydroelectric Power Generation -- 9.1.1. Principles of hydroelectric plant operation -- 9.1.2. Types of hydraulic turbine -- 9.1.2.1. Impulse turbines -- 9.1.2.2. Reaction turbines -- 9.1.2.3. Axial flow turbines -- 9.1.3. Pumped storage systems -- 9.1.4. Worked examples on hydroelectric power generation -- 9.2. Tidal Power Schemes -- 9.3. Methods of Utilizing Tidal Power -- 9.3.1. Tidal lagoons -- 9.3.2. Tidal stream farms -- 9.3.3. Tidal barrages. 9.3.3.1. Power available from a tidal barrage -- 9.3.3.2. Costs of tidal barrage schemes -- 9.3.3.3.Combination of a pumped storage facility with a tidal barrage scheme -- 9.3.3.4. Features of tidal barrage schemes -- 9.3.3.5. Worked examples on tidal barrage schemes -- 9.4. Wave Power -- 9.4.1. Basic properties of ideal deep-water waves -- 9.4.2. Power extractable from practical deep-water waves -- 9.4.3. Worked examples on wave energy -- 9.4.4. Types of wave power converters -- 9.4.5. Worked examples on wave-energy converter devices -- 9.4.6. Features of wave power systems: Summary -- 9.5. Problems and Review Questions -- References -- ch. 10 Wind Energy -- 10.1. Nature and Origin of the Wind -- 10.1.1. Atmospheric pressure -- 10.1.2. Atmospheric density -- 10.2. The Availability of Wind Supply -- 10.2.1. Global survey -- 10.2.2. Energy content of the wind -- 10.2.3. Wind energy supply in Europe -- 10.2.4. Wind energy supply in the USA. 10.3. Power Available in the Wind -- 10.3.1. Theoretical power available -- 10.3.2. Maximum theoretical power extractable from the wind -- 10.3.3. Practical power extractable from the wind -- 10.3.4. Tip-speed ratio -- 10.4. Mechanical Features of Wind Machines -- 10.4.1. Axial thrust (pressure) -- 10.4.2. The yaw"r; effect -- 10.4.3. Gyroscopic forces and vibrations -- 10.4.4. Centrifugal forces -- 10.4.5. Solidity factor -- 10.4.6. Two rotor blades or three rotor blades? -- 10.4.7. Shaft torque and power -- 10.5. Fixed Rotational Speed or Variable Rotational Speed? -- 10.5.1. Constant speed operation -- 10.5.2. Variable speed operation -- 10.6. Efficiency Considerations in Wind-Powered Turbine Systems -- 10.7. Wind Energy Turbines -- 10.7.1.19th-century windmills -- 10.7.2. Early 20th-century wind energy turbines -- 10.7.3. Later 20th-century wind energy turbines -- 10.8. Modern Large Wind Power Installations -- 10.8.1. Review of some installations. 10.8.2. Types of wind generator systems -- 10.8.2.1. Fixed-speed and directly coupled cage induction generator -- 10.8.2.2. Variable-speed, doubly fed induction generator -- 10.8.2.3. Variable-speed, direct drive synchronous generator -- 10.9. Vertical Axis Wind Machines -- 10.9.1. The Savonius design -- 10.9.2. The Darrieus design -- 10.9.3. Other forms of vertical axis machine -- 10.10. Worked Numerical Examples on Wind Turbine Operations -- 10.11. Problems and Review Questions -- References -- ch. 11 Solar Heating of Water or Air -- 11.1. Radiation from the Sun -- 11.2. Seasonal Variation of Solar Radiation -- 11.3. Classification of the Collection of Solar Energy -- 11.4. Solar Water Heating (Domestic) -- 11.4.1. Operation of a garden greenhouse -- 11.4.2. Solar flat-plate collectors -- 11.4.3.A typical domestic solar water heating system -- 11.4.4. Worked examples involving solar flat-plate collectors -- 11.5. Solar Water Heating (Industrial). 11.5.1. Solar tracking systems -- 11.5.1.1. Parabolic dish collector -- 11.5.1.2. Solar power tower -- 11.5.1.3. Linear focus collectors -- 11.5.2. Solar non-tracking systems -- 11.5.2.1. Evacuated tube collectors -- 11.5.2.2.Compound parabolic concentrator -- 11.5.3. Worked examples involving solar thermionic concentrator systems -- 11.6. Passive Solar Space Heating of Buildings -- 11.6.1. Direct gain solar systems -- 11.6.2. Indirect gain solar systems -- 11.6.2.1. Thermal storage wall -- 11.6.2.2. Solar greenhouse (sunspace) -- 11.6.2.3. Roof pond -- 11.6.2.4. Solar salt pond -- 11.7. Problems and Review Questions -- References --

Additional References -- ch. 12 Solar Photovoltaic Conversion -- 12.1. Basic Features of Solar Cells and Solar Systems -- 12.2. Physical Nature of Semiconductor Materials -- 12.2.1. Group-3 (acceptor) impurities -- 12.2.2. Group-5 (donor) impurities -- 12.3. Photovoltaic Materials -- 12.3.1. Crystalline silicon (c--Si). 12.3.2. Amorphous (uncrystalline) silicon (a--Si) -- 12.3.3. Materials other than silicon -- 12.4. Cost of Solar Photovoltaic Electricity -- 12.5. Operation of the Semiconductor Diode and Solar Photovoltaic Cell -- 12.6. Physical Properties of the Solar Photovoltaic Cell -- 12.7. Electrical Output Properties of the Solar Photovoltaic Cell -- 12.7.1. Maximum power delivery -- 12.7.2. Equivalent circuits -- 12.7.3. Load lines in the current--voltage plane -- 12.7.4. Arrays of solar photovoltaic cells -- 12.7.5. Effect of temperature on solar cell operation -- 12.8. Applications of Photovoltaic Cells -- 12.9. The Future Challenge for Photovoltaics -- 12.10. Worked Examples -- 12.11. Problems and Review Questions -- References -- ch. 13 Biological Energy and Chemical Energy -- 13.1. Biomass and Biofuels -- 13.1.1. Natural vegetation -- 13.1.2. Energy tree plantations -- 13.1.3. Specific energy crops -- 13.1.4. Use of wastes -- 13.1.5. Water-based biomass -- 13.2. Photosynthesis. 13.3. Methods of Industrial Biomass Conversion -- 13.3.1.Combustion -- 13.3.2. Pyrolysis -- 13.3.3. Gasification of biomass -- 13.3.4. Liquid and gaseous fuels from biomass -- 13.3.4.1. Chemical reduction -- 13.3.4.2. Alcoholic fermentation (ethanol) -- 13.3.4.3. Anaerobic digestion to produce biogas -- 13.4.

	Wood as a Fuel -- 13.5. Energy from Wastes -- 13.5.1. Solid waste disposal in landfill sites -- 13.5.2. Solid waste disposal using municipal incinerators (combustors) -- 13.5.3. Worked examples on solid waste incineration -- 13.5.4. Liquid and gaseous wastes -- 13.6. The Fuel Cell -- 13.7. Problems and Review Questions -- References -- ch. 14 The Energy Future -- 14.1. The Energy Problems -- 14.2. An Energy Strategy -- 14.3. The Long-Term Energy Future -- 14.3.1. Nuclear fission using breeder reactors -- 14.3.2. Solar energy -- 14.3.3. Controlled thermonuclear fusion -- 14.3.4. Geothermal energy -- 14.4. What Shall We Do When the Oil Runs Out? -- References.
Subjects	Power resources. Energy conversion. Energy consumption. Renewable energy sources.
Notes	Previous edition: 2003. Includes bibliographical references and index.

Environmental review of the proposed 130 mw generating facility within the Cove Point liquefaction project: prepared for Maryland Department of Natural Resources Power Plant Research Program

LCCN	2014489598
Type of material	Book
Personal name	Mountain, Diane, compiler.
Main title	Environmental review of the proposed 130 mw generating facility within the Cove Point liquefaction project: prepared for Maryland Department of Natural Resources Power Plant Research Program / prepared by D. Mountain, M. Johnson, R. Keating, Environmental Resources

	Management, Inc., Annapolis, MD; M. DiPrinzio, J. Flannery, M. Garrison, J. Ross, Environmental Resources Management, Inc., Malvern, PA; S. Harriott, Versar, Inc. ESM OperationsColumbia, MD; P. Hall, Metametrics, Inc., Charlottesville, VA; M. Brown, Spectrum Environmental Sciences, Inc., Frederick, MD
Published/Produced	[Annapolis, MD]: [Maryland Department of Natural Resources], [2014]
Description	vii, 237 pages: illustrations, maps; 28 cm
LC classification	TP761.L5 M68 2014
Related names	Johnson, M., compiler. Keating, Robert, compiler. DiPrinzio, Mark, compiler. Flannery, J, compiler. Garrison, Mark, compiler. Ross, Julie, 1959- compiler. Harriott, Steve, compiler. Hall, Peter D., compiler. Brown, M., compiler. Maryland. Department of Natural Resources. Maryland Power Plant Research Program. ERM, Inc. Versar, Inc. MetaMetrics Inc. Spectrum Environmental Sciences, Inc.
Summary	This document summarizes the State of Maryland's evaluation of the proposed Dominion Cove Point Generating Facility in Calvert County, Maryland. The project received final approval from the Public Service Commission (PSC) on 30 May 2014.
Subjects	Liquefied natural gas--Storage--Safety measures. Liquefied natural gas--Transportation--Safety measures.

	Marine terminals--Maryland--Cove Point.
Notes	"December 2014"
	"DNR Publication No. 12-2102015-750" -- Cover
	"PPSE-CPT-02" -- Cover
	Includes bibliographical references and appendixes.

Federal financial incentives to induce early experience producing unconventional liquid fuels

LCCN	2008053726
Type of material	Book
Personal name	Camm, Frank A., 1949-
Main title	Federal financial incentives to induce early experience producing unconventional liquid fuels / Frank Camm, James T. Bartis, Charles J. Bushman.
Published/Created	Santa Monica, CA: RAND Corp., 2008.
Description	xvii, 77 p.: ill. (some col.); 28 cm.
ISBN	9780833045102 (pbk.: alk. paper)
	0833045105 (pbk.: alk. paper)
LC classification	TP352 .C36 2008
Related names	Bartis, James T., 1945-
	Bushman, Claudia L.
Contents	Introduction -- Designing an effective long-term public-private relationship -- Assessing financial effects under uncertainty --Policy effects with 100-percent equity financing -- Policy effects with debt financing -- Implications for robust financial-incentive packages -- Can formal source selection help the government create an integrated policy? -- Conclusions -- Appendixes: A. Structure of the spreadsheet analysis that implements the cash-flow model -- B. How debt and loan guarantees affect investors and the government.
Subjects	Coal liquefaction--United States.

	Federal aid to energy development--United States.
	Energy development--United States.
Notes	Includes bibliographical references (p. 75-77).
Series	Technical report; TR-586-AF/NETL
	Technical report (Rand Corporation); TR-586-AF/NETL.

Feedstock recycling and pyrolysis of waste plastics: converting waste plastics into diesel and other fuels

LCCN	2005029094
Type of material	Book
Main title	Feedstock recycling and pyrolysis of waste plastics: converting waste plastics into diesel and other fuels / edited by John Scheirs and Walter Kaminsky.
Published/Created	Chichester, UK; Hoboken, NJ: J. Wiley & Sons, c2006.
Description	xxvii, 785 p.: ill.; 26 cm.
Links	Table of contents http://www.loc.gov/catdir/toc/ecip061/2005029094.html
	Publisher description http://www.loc.gov/catdir/enhancements/fy0644/2005029094-d.html
	Contributor biographical information http://www.loc.gov/catdir/enhancements/fy0740/2005029094-b.html
ISBN	0470021527 (cloth: alk. paper)
	9780470021521 (cloth: alk. paper)
LC classification	TD899.P55 F45 2006
Related names	Scheirs, John.
	Kaminsky, W. (Walter), 1941-
Contents	Introduction to feedstock recycling of plastics / A. Buekens -- Acid-catalyzed cracking of polyolefins: primary reaction mechanisms / Robert L. White -- Catalytic upgrading of plastic wastes / J. Aguado,

D.P. Serrano, J.M. Escola -- Thermal and catalytic conversion of polyolefins / Jerzy Walendziewski -- Thermal and catalytic degradation of waste hdpe / Kyong-Hwan Lee -- Development of a process for the continuous conversion of waste plastics mixtures to fuel / Takao Masuda, Teruoki Tago -- Catalytic degradation of plastic waste to fuel over microporous materials / George Manos -- Liquefaction of municipal waste plastics over acidic and nonacidic catalysts / Jale Yanik, Tamer Karayildirim -- Kinetic model of the chemical and catalytic recycling of waste polyethylene into fuels / Norbert Miskolczi -- Production of gaseous and liquid fuels by pyrolysis and gasification of plastics: technological approach / C. Gisèle Jung, André Fontana -- Yield and composition of gases and oils/waxes from the feedstock recycling of waste plastic / Paul T. Williams -- Composition of liquid fuels derived from the pyrolysis of plastics / Marianne Blazsó -- Production of premium oil products from waste plastic by pyrolysis and hydroprocessing / S.J. Miller, N. Shah, G.P. Huffman -- The conversion of waste plastics/petroleum residue mixtures to transportation fuels / Mohammad Farhat Ali, Mohammad Nahid Siddiqui -- Overview of commercial pyrolysis processes for waste plastics / John Scheirs -- Fluidized bed pyrolysis of plastic wastes / Umberto Arena, Maria Laura Mastellone -- The Hamburg fluidized-bed pyrolysis process to recycle polymer wastes and tires / Walter Kaminsky -- Liquefaction of PVC mixed plastics / Thallada Bhaskar, Yusaku Sakata -- Liquid fuel from plastic wastes using extrusion-rotary kiln reactors / Sam Behzadi, Mohammed Farid -- Rotary kiln pyrolysis of polymers containing heteroatoms / Andreas Hornung, Helmut Seifert -- Microwave pyrolysis of plastic wastes / C. Ludlow-Palafox, H.A. Chase -- Continuous thermal process for cracking polyolefin wastes to produce hydrocarbons / Jean Dispons -- Waste plastic pyrolysis in free-fall reactors / Ali Y. Bilgesü, M. Çetin Koçak, Ali Karaduman -- Monomer recovery of plastic waste in a fluidized bed process / Walter Kaminsky -- Feedstock recycling of PET / Toshiaki Yoshioka, Guido Grause -- The liquefaction of plastic containers and packaging in Japan / A. Okuwaki, T. Yoshioka, M. Asai, H. Tachibana, K. Wakai, K. Tada -- Process and equipment for conversions of waste plastics into fuels / Alka Zadgaonkar -- Converting waste plastics into liquid fuel by pyrolysis: developments in China / Yuan Xingzhong.

Subjects	Plastics industry and trade--Waste minimization. Plastics--Recycling. Pyrolysis. Plastic scrap--Recycling.
Notes	Includes bibliographical references and index.
Series	Wiley series in polymer science

Final environmental impact statement, Cameron Liquefaction Project.

LCCN	2014406369
Type of material	Book
Main title	Final environmental impact statement, Cameron Liquefaction Project.

Published/Produced	Washington, District of Columbia: Federal Energy Regulatory Commission, Office of Energy Projects, 2014.
Description	1 volume (various pagings): illustrations, maps, charts; 29 cm
LC classification	TP350 .F56 2014
Portion of title	Cameron Liquefaction Project
Related names	United States. Federal Energy Regulatory Commission. Office of Energy Projects, issuing body. United States. Army. Corps of Engineers. United States. Coast Guard. United States. Department of Energy. United States. Department of Transportation.
Subjects	Liquefied natural gas--Environmental aspects--Louisiana--Hackberry. Exports--United States--Planning. Environmental impact statements--Louisiana--Hackberry.
Notes	Cover title. "April 2014." Format not distributed to depository libraries. "Cooperating agencies: US. Army Corps of Engineers; U.S. Coast Guard; U.S. Department of Energy (DOE Docket no. FE 11-162-LNG); U.S. Department of Transportation." "Docket nos.: CP13-25-000; CP13-27-000." "FERC/EIS-248F." "DOE/EIS-0488." Includes bibliographical references.
Additional formats	Online version: Final environmental impact statement,Cameron Liquefaction Project (OCoLC) 880178397

Fischer-Tropsch refining

LCCN	2011534447
Type of material	Book
Personal name	Klerk, Arno de.
Main title	Fischer-Tropsch refining / Arno de Klerk.
Edition	[1st. ed.]
Published/Created	Weinheim, Germany: Wiley-VCH, c2011.
Description	xxi, 620 p.: ill., map; 25 cm.
ISBN	9783527635603 (oBook) 9783527635610 (ePub) 9783527326051 (hbk.: alk. paper) 3527326057 (hbk.: alk. paper)
LC classification	TP355 K55 2011
Contents	Introduction. Fischer-Tropsch facilities at a glance -- Refining and refineries at a glance -- Production of Fischer-Tropsch syncrude. Synthesis gas production, cleaning, and conditioning -- Fischer-Tropsch synthesis -- Fischer-Tropsch gas loop -- Industrial Fischer-Tropsch facilities. German Fischer-Tropsch facilities -- American hydrocol facility -- Sasol 1 facility -- Sasol 2 and 3 facilities -- Mossgas facility -- Shell middle distillate synthesis (SMDS) facilities -- Oryx and Escravos gas-to-liquids facilities -- Synthetic transportation fuels. Motor-gasoline -- Jet fuel -- Diesel fuel -- Refining technology. Refining technology selection -- Dehydration, etherification, and hydration -- Isomerization -- Oligomerization -- Aromatic alkylation -- Cracking -- Reforming and aromatization -- Chemical technologies -- Refinery design. Principles of refinery design -- Motor-gasoline refining -- Jet fuel refining -- Diesel fuel refining -- Chemicals and lubricant refining.
Subjects	Fischer-Tropsch process.

	Gases--Liquefaction.
	Synthetic fuels--Refining.
Notes	Includes bibliographical references and index.

Freeport LNG Liquefaction Project Phase II Modification Project: draft enviromental impact statement.

LCCN	2014406525
Type of material	Book
Main title	Freeport LNG Liquefaction Project Phase II Modification Project: draft enviromental impact statement.
Published/Produced	Washington, DC: Federal Energy Regulatory Commission, Office of Energy Projects, [2014]
Description	1 volume (various pagings); 21 cm
LC classification	TD195.G3 F735 2014
Related names	United States. Federal Energy Regulatory Commission. Office of Energy Projects, issuing body.
Subjects	Freeport LNG Development, L.P.
	Liquefied natural gas pipelines--Environmental aspects--Texas--Brazoria County.
	Liquefied natural gas shipping terminals--Environmental aspects--Texas--Brazoria County.
	Environmental impact statements--Texas--Brazoria County.
Notes	Title from cover.
	"March 2014."
	"Freeport LNG Development, L.P; FLNG Liquefaction, LLC; FLNG Liquefaction 2, LLC; FLNG Liquefaction 3, LLC."
	"Docket Nos. CP12-509-000; CP12-29-000; PF11-2-000; FERC/EIS-0250D; DOE/EIS-0487."
	Includes bibliographical references.

Future of federal coal: status, availability, and impact of technological advances in using coal to create alternative energy resources: oversight hearing before the Subcommittee on Energy and Mineral Resources of the Committee on Resources, U.S. House of Representatives, One Hundred Ninth Congress, second session, Thursday, May 4, 2006.

LCCN	2006460970
Type of material	Book
Corporate name	United States. Congress. House. Committee on Resources. Subcommittee on Energy and Mineral Resources.
Main title	Future of federal coal: status, availability, and impact of technological advances in using coal to create alternative energy resources: oversight hearing before the Subcommittee on Energy and Mineral Resources of the Committee on Resources, U.S. House of Representatives, One Hundred Ninth Congress, second session, Thursday, May 4, 2006.
Published/Created	Washington: U.S. G.P.O.: For sale by the Supt. of Docs., U.S. G.P.O., 2006.
Description	iii, 89 p.: ill.; 24 cm.
Links	http://purl.access.gpo.gov/GPO/LPS73896
ISBN	016076856X
LC classification	KF27 .R3846 2006
Subjects	Coal liquefaction--Technological innovations--United States. Synthetic fuels--Technological innovations--United States. Fischer-Tropsch process.
Notes	Distributed to some depository libraries in microfiche. Shipping list no.: 2006-0315-P. "Serial no. 109-51."

	Includes bibliographical references.
Reproduction no./Source	552-070-34595-7 GPO
Additional formats	Also available via Internet from the GPO Access web site. Address as of 9/12/06: http://frwebgate.access.gpo.gov/cgi-bin/getdoc.cgi?dbname=109%5Fhouse%5Fhearings&docid=f:27378.pdf; current access available via PURL.

Gasoline, diesel, and ethanol biofuels from grasses and plants

LCCN	2009042276
Type of material	Book
Personal name	Gupta, Ram B.
Main title	Gasoline, diesel, and ethanol biofuels from grasses and plants / Ram B. Gupta, Ayhan Demirbas.
Published/Created	New York: Cambridge University Press, 2010.
Description	xiv, 230 p.: ill., maps; 26 cm.
ISBN	9780521763998 (hardback) 0521763991 (hardback)
LC classification	TP248.27.P55 G87 2010
Related names	Demirbas, Ayhan.
Summary	"This book introduces readers to second-generation biofuels obtained from non-food biomass, such as forest residue, agricultural residue, switch grass, and corn stover"--Provided by publisher. "The world is currently faced with two significant problems: fossil fuel depletion and environmental degradation, which are continuously being exacerbated due to increasing global energy consumption. As a substitute for petroleum, renewable fuels have been receiving increasing attention due a variety of environmental, economic, and societal benefits. The first-

generation biofuels - ethanol from sugar or corn and biodiesel from vegetable oils - are already on the market. The goal of this book is to introduce readers to second-generation biofuels obtained from non-food biomass, such as forest residue, agricultural residue, switch grass, corn stover, waste wood, municipal solid wastes, and so on. Various technologies are discussed, including cellulosic ethanol, biomass gasification, synthesis of diesel and gasoline, bio-crude by hydrothermal liquefaction, bio-oil by fast pyrolysis, and the upgradation of biofuel. This book strives to serve as a comprehensive document presenting various technological pathways and environmental and economic issues related to biofuels"--Provided by publisher.

Contents 1. Introduction -- 2. Air pollution and global warming from the use of fossil fuels -- 3. Renewable energy sources -- 4. Biomass availability in the world -- 5. Convention ethanol production from corn and sugarcane -- 6. Ethanol from biomass by fermentation -- 7. Biodiesel from vegetable oils -- 8. Diesel from biomass gasification followed by Fischer-Tropsch synthesis -- 9. Bio-oil from biomass pyrolysis -- 10. Bio-crude from biomass hydrothermal liquefaction -- 11. Solar and wind energy for biofuel production -- 12. Environmental impacts of biofuels -- 13. Economic impact of biofuels -- 14. Biofuel policy.

Subjects Plant biomass.
Forest biomass.
Biomass energy.

Notes Includes bibliographical references and index.

Gas-to-liquids processes for chemicals and energy production

LCCN	2001276648
Type of material	Book
Personal name	Gobina, Edward.
Main title	Gas-to-liquids processes for chemicals and energy production / Edward Gobina.
Published/Created	Norwalk, Conn. (25 Van Zant St., Norwalk 06855): Business Communications, c2000.
Description	xxii, 239 p., [3] leaves: ill.; 28 cm.
ISBN	1569651833
LC classification	HD9581.2.L573 U642 2000
Related names	Business Communications Co.
Subjects	Liquefied natural gas industry--United States. Liquefied petroleum gas industry--United States. Liquefied gases. Gases--Liquefaction. Market surveys--United States.
Notes	"October 2000"--T.p. verso. Includes bibliographical references (leaf xviii-xx).
Series	Business opportunity report; E-101

Geologic evaluation and hazard potential of liquefaction-induced landslides along the Wasatch Front, Utah

LCCN	2003628224
Type of material	Book
Personal name	Harty, Kimm M.
Main title	Geologic evaluation and hazard potential of liquefaction-induced landslides along the Wasatch Front, Utah / by Kimm M. Harty and Mike Lowe.
Published/Created	Salt Lake City, UT: Utah Geological Survey, 2003.
Description	40 p.: ill., maps; 28 cm., + 16 plates in pocket.
ISBN	155791687X
LC classification	QE599.2 .H37 2003
Related names	Lowe, Mike.

Subjects	Landslide hazard analysis--Wasatch Range (Utah and Idaho). Slopes (soil mechanics)--Wasatch Range (Utah and Idaho). Soil liquefaction--Utah.
Notes	Includes bibliographical references (p. 30-32).
Series	Special study / Utah Geological Survey; 104 Special study (Utah Geological Survey); 104.

Geomechanics of failures

LCCN	2010925324
Type of material	Book
Personal name	Puzrin, A. M. (Alexander M.)
Main title	Geomechanics of failures / A.M. Puzrin, E.E. Alonso, N.M. Pinyol.
Published/Created	Dordrecht; New York: Springer, c2010.
Description	vii, 245 p.: ill. (some col.); 24 cm.
ISBN	9789048135301 (hbk.: acid-free paper) 9048135303 (hbk.: acid-free paper) 9789048135318 9048135311
Publisher no.	12695737
LC classification	TA656 .P87 2010
Related names	Alonso, Eduardo E., 1947- Pinyol, N. M. (Núria M.)
Contents	Interaction between neighboring structures: Mexico City Metropolitan Cathedral, Mexico -- Unexpected excessive settlements: Kansai International Airport, Japan -- Bearing capacity failure: the Tower of Pisa, Italy -- Bearing capacity failure: Transcona Grain Elevator, Canada -- Caisson failure induced by liquefaction: Barcelona Harbour, Spain -- Braced excavation collapse: Nicoll Highway, Singapore -- Tunnel excavation

	collapse: Barrás Square, Spain -- Tunnel face instability: Floresta Tunnels, Spain.
Subjects	Structural failures--Investigation--Case studies. Structural analysis (Engineering)--Mathematics. Soil mechanics--Mathematics. Geomechanik. Ingenieurgeologie. Instabilität.
Notes	Includes bibliographical references.

Geotechnical earthquake engineering handbook: with the 2012 International building code

LCCN	2012022486
Type of material	Book
Personal name	Day, Robert W.
Main title	Geotechnical earthquake engineering handbook: with the 2012 International building code / Robert W. Day.
Edition	Second edition.
Published/Produced	New York: McGraw-Hill, [2012] ©2012
Description	1 volume (various pagings): illustrations; 24 cm
ISBN	9780071792387 (hardback: acid-free paper)
LC classification	TA654.6 .D39 2012
Related titles	International building code.
Summary	"The latest methods for designing seismically sound structuresFully updated for the 2012 International Building Code, Geotechnical Earthquake Engineering Handbook, Second Edition discusses basic earthquake principles, common earthquake effects, and typical structural damage caused by seismic shaking. Earthquake computations for conditions commonly encountered by design engineers, such as

liquefaction, settlement, bearing capacity, and slope stability, are included. Site improvement methods that can be used to mitigate the effects of earthquakes on structures are also described in this practical, comprehensive guide.Coverage includes: Basic earthquake principles Common earthquake effects Earthquake structural damage Site investigation for geotechnical earthquake engineering Liquefaction Earthquake-induced settlement Bearing capacity analyses for earthquakes Slope stability analyses for earthquakes Retaining wall analyses for earthquakes Other geotechnical earthquake engineering analyses Grading and other soil improvement methods Foundation alternatives to mitigate earthquake effects Earthquake provisions in building codes "-- Provided by publisher.

"This one-stop resource--filled with in-depth earthquake engineering analysis, testing procedures, seismic and construction codes--features new coverage of the 2012 International Building Code"-- Provided by publisher.

Subjects	Earthquake engineering--Handbooks, manuals, etc.
	Engineering geology--Handbooks, manuals, etc.
	Technology & Engineering / Civil / Earthquake.
Notes	Includes bibliographical references and index.

Grounded!: amazing classroom demonstrations in soil mechanics

LCCN	2015000859
Type of material	Book
Personal name	Elton, David J., author.
Main title	Grounded!: amazing classroom demonstrations in soil mechanics / David J. Elton, Ph.D., P.E.

Published/Produced	Reston, Virginia: American Society of Civil Engineers, [2015]
Description	xiii, 191 pages; 23 cm
ISBN	9780784413920 (print: alk. paper)
LC classification	TA710 .E459 2015
Summary	"Elton presents 35 serious but entertaining experiments for budding scientists and engineering students that teach the fundamentals of soil mechanics and illustrate the dynamics of how soils behave and how they can be manipulated."--Provided by publisher.
Contents	Exploding soils -- Rice is nice: an exercise in pile capacity -- Break that block: the power of bentonite -- A retaining wall made of: paper! -- Slip slidin' away with effective stress -- Magic beaker: sand defies gravity -- Effective potato chips -- Foam balls in a pipe: soil arching -- Geotechnical Rorschach test (water flows around a drain, instead of into the drain) -- Hairy soil -- Sand cylinder hoisting -- Water that won't go through a sieve; water that does -- Dilation in someone's hand: (silty soil identification) -- Piping under a sheetpile dam -- Jar of rocks: soil void ratio, porosity, unit weight and water content of soils -- Liquefaction by upward gradient -- Blow it out: critical hydraulic gradient. flowlines -- Relative density of soil in a graduated cylinder -- Soil filter demonstration using marbles -- Sand in a jar, strong as you are -- Slinky shear waves (and compression ones, too!) -- Soil bridge in a tube -- Capillarity and cotton balls, plus (bonus!) non-woven geotextile strength demonstration -- Stick slip behavior -- Tilt me up Scotty: interface friction -- Soils relaxing: the angle of repose -- Falling head over heels for

	permeability: the permeability of ooblech -- Bulking of soils: aka, less soil for your money -- Geotextile puncture test -- Soil dilation and compaction using ping pong balls -- Campus geology tour -- Settlement rates of soils in jars – Bearing capacity: bigger is better -- Retaining wall in a box -- Rockfall simulator lab.
Subjects	Soils--Experiments. Soil mechanics--Experiments.
Notes	Includes bibliographical references (pages 187) and index.

Guidance document, conducting paleoliquefaction studies for earthquake source characterization

LCCN	2017492360
Type of material	Book
Personal name	Tuttle, M., author.
Main title	Guidance document, conducting paleoliquefaction studies for earthquake source characterization / prepared: by M. Tuttle [and four others].
Published/Produced	Washington, DC: United States Nuclear Regulatory Commission, Office of Nuclear Reactor Regulation, 2018.
Description	1 volume (various pagings): color illustrations; 28 cm
LC classification	QE539.2.S34 T87 2018
Variant title	Conducting paleoliquefaction studies for earthquake source characterization
Related names	U.S. Nuclear Regulatory Commission. Office of Nuclear Regulatory Research. Division of Engineering Technology.
Subjects	Earthquake hazard analysis. Soil liquefaction.
Notes	"Date published: January 2018."

"NUREG/CR-7238."
Includes bibliographical references.

Hazard analysis of seismic soil liquefaction.

LCCN 2017935832
Type of material Book
Main title Hazard analysis of seismic soil liquefaction.
Published/Produced New York, NY: Springer Berlin Heidelberg, 2017.
ISBN 9789811043789

Implementation of the provisions of the Energy Policy Act of 2005: hearings before the Energy and Natural Resources, One Hundred Ninth Congress, second session.

LCCN 2006470096
Type of material Book
Corporate name United States. Congress. Senate. Committee on Energy and Natural Resources.
Main title Implementation of the provisions of the Energy Policy Act of 2005: hearings before the Energy and Natural Resources, One Hundred Ninth Congress, second session.
Published/Created Washington: U.S. G.P.O.: For sale by the Supt. of Docs., U.S. G.P.O., 2006-
Description v. <1-3 >; 24 cm.
Links Pt. 1, text version http://purl.access.gpo.gov/GPO/LPS74580
Pt. 1, PDF version http://purl.access.gpo.gov/GPO/LPS74581
Pt. 2, text version http://purl.access.gpo.gov/GPO/LPS75088
Pt. 2, PDF version http://purl.access.gpo.gov/GPO/LPS75089
Pt. 3, text version http://purl.access.gpo.gov/GPO/LPS74977

	Pt. 3, PDF version http://purl.access.gpo.gov/GPO/LPS74978
ISBN	0160770327 (pt.1) 0160772389 (pt.2) 016077201X (pt.3)
LC classification	KF26 .E55 2006n
Subjects	Coal liquefaction--Technological innovations--United States. Coal gasification--Technological innovations--United States. Hydroelectric power plants--Licenses--United States. Energy development--Law and legislation--United States.
Notes	"On coal liquefaction technology, coal gasification technology, and licensing of hydroelectric facilities"--Pt. 1. "On electricity reliability provisions, nuclear power provisions, next generation nuclear plant, and renewable fuel standard and the future potential of biofuels"--Pt. 2. "On enhancing oil and gas production, geothermal energy and other renewables, and hydrogen and fuel cell research and development"--Pt. 3. "April 24, 2006, May 1, 2006, May 8, 2006"--Pt. 1. "May 15, 2006, May 22, 2006, June 12, 2006, June 19, 2006"--Pt. 2. "June 27, 2006, July 11, 2006, July 17, 2006"--Pt. 3. "Printed for the use of the Committee on Energy and Natural Resources." Distributed to some depository libraries in microfiche.

	Shipping list no.: 2006-0327-P (pt. 1), 2007-0024-P (pt. 2), 2007-0016-P (pt. 3).
	Includes bibliographical references.
Reproduction no./Source	552-070-34670-8 (pt.1) GPO
	552-070-34811-5 (pt.2) GPO
	552-070-34788-7 (pt.3) GPO
Additional formats	Also available via Internet from the GPO Access web site. Addresses as of 10/19/06: http://frwebgate.access.gpo.gov/cgi-bin/getdoc.cgi?dbname=109%5Fsenate%5Fhearings&docid=f:28983.wais (pt. 1, text version), http://frwebgate.access.gpo.gov/cgi-bin/getdoc.cgi?dbname=109%5Fsenate%5Fhearings&docid=f:28983.pdf (pt. 1, PDF version), http://frwebgate.access.gpo.gov/cgi-bin/getdoc.cgi?dbname=109%5Fsenate%5Fhearings&docid=f:29644.wais (pt. 2, text version), http://frwebgate.access.gpo.gov/cgi-bin/getdoc.cgi?dbname=109%5Fsenate%5Fhearings&docid=f:29644.pdf (pt. 2, PDF version), http://frwebgate.access.gpo.gov/cgi-bin/getdoc.cgi?dbname=109%5Fsenate%5Fhearings&docid=f:30004.wais (pt. 3, text version), http://frwebgate.access.gpo.gov/cgi-bin/getdoc.cgi?dbname=109%5Fsenate%5Fhearings&docid=f:30004.pdf (pt. 3, PDF version); current access available via PURLs.
Series	S. hrg.; 109-503
	United States. Congress. Senate. S. hrg.; 109-503.

Lake Charles Liquefaction Project: draft environmental impact statement.

LCCN	2014489581
Type of material	Book

Main title	Lake Charles Liquefaction Project: draft environmental impact statement.
Published/Produced	Washington, DC: Federal Energy Regulatory Commission, Office of Energy Project, [2015]
Description	1 volume (various pagings); 28 cm
LC classification	TD195.G3 L346 2015
Related names	United States. Federal Energy Regulatory Commission. Office of Energy Projects.
Subjects	Liquefied natural gas shipping terminals--Environmental aspects--Louisiana--Lake Charles. Liquefied natural gas pipelines--Environmental aspects--Louisiana--Lake Charles. Environmental impact statements--Louisiana--Lake Charles.
Notes	Title from cover. "April 2015"--Spine. Includes bibliographical references.

Lake Charles Liquefaction Project: final environmental impact statement.

LCCN	2014496977
Type of material	Book
Corporate name	United States. Federal Energy Regulatory Commission. Office of Energy Projects, author.
Main title	Lake Charles Liquefaction Project: final environmental impact statement.
Published/Produced	Washington, DC: Federal Energy Regulatory Commission, Office of Energy Project, [2015]
Description	1 volume (various pagings): color illustrations; 28 cm
LC classification	TD195.G3 U548 2015
Subjects	Liquefied natural gas shipping terminals--Environmental aspects--Louisiana--Lake Charles.

	Liquefied natural gas pipelines--Environmental aspects--Louisiana--Lake Charles.
	Environmental impact statements--Louisiana--Lake Charles.
Notes	Title from cover.
	"Cooperating agencies: U.S. Department of Energy, U.S. Coast Guard, U.S. Fish and Wildlife Service, U.S. Army Corps of Engineers, U.S. Department of Transportation."
	"Trunkline Gas Company, LLC, Lake Charles LNG Company, LLC, and Lake Charles LNG Export Company, LLC."
	"FERC docket nos. CP14-119-000, CP14-120-000, and CP14-122-000. DOE docket nos. 11-59-LNG and 13-04-LNG."
	"FERC/EIS-0258F, DOE/EIS-0491."
	"August 2015"--Spine.
	Format not distributed to depository libraries.
	Includes bibliographical references.
Additional formats	Online version: United States. Federal Energy Regulatory Commission. Office of Energy Projects. Lake Charles Liquefaction Project. (OCoLC)936564069

Landslides: mass wasting, soil, and mineral hazards

LCCN	2007024674
Type of material	Book
Personal name	Kusky, Timothy M.
Main title	Landslides: mass wasting, soil, and mineral hazards / Timothy Kusky.
Published/Created	New York: Facts on File, c2008.
Description	xvi, 128 p.: ill., maps; 25 cm.
Links	Table of contents only http://www.loc.gov/catdir/toc/ecip0720/2007024674.html

ISBN	9780816064656 (hc: acid-free paper) 0816064652 (hc: acid-free paper)
LC classification	QE599.A2 .K87 2008
Summary	"Examines the formation of soil from regolith--the outermost layer of Earth--and bedrock and discusses some of the hazardous elements that are concentrated in it"--P. [4] of cover.
Contents	Weathering and the formation of soils -- Hazardous elements in soils and the regolith -- Water in the regolith: expansive clays, liquefaction, and groundwater pollution -- Mass wasting processes -- Undersea landslides -- Landslide disasters -- Reduction of landslide hazards and damages.
Subjects	Landslides--Juvenile literature. Mass-wasting--Juvenile literature. Landslide hazard analysis--Juvenile literature. Landslides.
Notes	Includes bibliographical references (p. 113-120) and index. 012 up.
Series	The hazardous Earth Hazardous earth.

Liquefaction and other ground failures in Imperial County, California, from the April 4, 2010, El Mayor-Cucapah earthquake

LCCN	2011506710
Type of material	Book
Main title	Liquefaction and other ground failures in Imperial County, California, from the April 4, 2010, El Mayor-Cucapah earthquake / by Timothy P. McCrink ... [et al.]
Published/Created	[Sacramento, Calif.]: California Dept. of Conservation, California Geological Survey;

	Reston, VA: U.S. Dept. of the Interior, U.S. Geological Survey, 2011.
Description	x, 84 p.: col. ill., col. maps; 28 cm.
LC classification	QE535.2.C2 L57 2011
Related names	McCrink, Timothy P.
	California Geological Survey.
	Geological Survey (U.S.)
Subjects	Earthquake hazard analysis--California.
	Earthquakes--California--Imperial County.
Notes	Title from cover.
	"Version 1.0."
	Maps in pocket inside back cover.
	Errata slip inserted.
	Includes bibliographical references (p. 80-81).
Additional formats	Also available online.
	Online version: Liquefaction and other ground failures in Imperial County, California, from the April 4, 2010, El Mayor-Cucapah earthquake. [Sacramento, Calif.]: California Dept. of Conservation, California Geological Survey; Reston, VA: U.S. Dept. of the Interior, U.S. Geological Survey, 2011 (OCoLC)773727524
Series	CGS Special report; 220
	USGS open file report; 2011-1071
	Special report (California Geological Survey); 220.
	U.S. Geological Survey open-file report; 2011-1071.

Liquefaction around marine structures

LCCN	2014412870
Type of material	Book
Personal name	Sumer, B. Mutlu, author.

Main title	Liquefaction around marine structures / B. Mutlu Sumer, Technical University of Denmark, Denmark.
Published/Produced	Singapore; New Jersey: World Scientific, [2014] ©2014
Description	xv, 453 pages: illustrations, maps; 26 cm 1 CD-ROM (4 3/4 in.)
ISBN	9789814329316 (hardcover) 9814329312 (hardcover)
LC classification	TA710 .S8462 2014
Contents	Introduction and physics of liquefaction -- Biot equations and solutions -- Residual liquefaction -- Momentary liquefaction -- Floatation of buried pipelines -- Sinking of pipelines and marine objects -- Liquefaction under standing waves -- Liquefaction at gravity structures -- Stability of rock berms in liquefied soil -- Impact of seismic-induced liquefaction -- Counter measures -- Small amplitude, linear waves -- Soil properties -- In situ relative density -- Hsu & Jeng coefficients -- List of symbols -- Author index -- Subject index. Contents of CD: Readme -- Video 1: Liquefaction under progressive waves, Part 1 -- Video 2: Liquefaction under progressive waves, Part 2 -- Video 3: Compaction stage of liquefaction/compaction sequence under progressive waves -- Video 4: Liquefaction understanding waves -- Video 5: Floatation of a buried pipeline under progressive waves.
Subjects	Soil liquefaction. Offshore structures.
Notes	Includes bibliographical references and indexes.
Series	Advanced series on ocean engineering, 1793-074X; vol. 39

	Advanced series on ocean engineering; v. 39. 1793-074X

Liquid natural gas in the United States: a history

LCCN	2014019903
Type of material	Book
Personal name	Hrastar, John.
Main title	Liquid natural gas in the United States: a history / John Hrastar.
Published/Produced	Jefferson, North Carolina: McFarland & Company, Inc., Publishers, [2014]
Description	vii, 277 pages: illustrations; 26 cm
ISBN	9780786478590 (softcover)
LC classification	TP761.L5 H73 2014
Summary	"When natural gas was discovered in Appalachia in the 19th century, its development and growth as a fuel was rapid. A 1944 disaster in Cleveland, when an LNG tank ruptured and caused a fire with 130 fatalities, caused the industry to pause. In 1959 a shipload of LNG went from Louisiana to Britain and restarted the LNG industry"-- Provided by publisher.
Contents	Inferno -- Discovery and growth -- Growth to shortage -- Discovery of gas liquefaction -- Liquid storage -- Setback -- Recovery -- Breakout: moving the energy.
Subjects	Liquefied natural gas--United States--History. Gas as fuel--United States--History.
Notes	Includes bibliographical references and index.

LNG: a nontechnical guide

LCCN	2007013934
Type of material	Book
Personal name	Tusiani, Michael D.

Main title	LNG: a nontechnical guide / Michael D. Tusiani and Gordon Shearer.
Published/Created	Tulsa, Okla.: PennWell, c2007.
Description	xxii, 436 p.: ill.; 24 cm.
Links	Table of contents only http://www.loc.gov/catdir/toc/ecip0715/2007013934.htm
ISBN	9780878148851 (hardcover) 087814885X (hardcover)
LC classification	HD9581.2.L572 T87 2007
Related names	Shearer, Gordon.
Contents	The LNG industry -- The evolution of the global gas and LNG markets -- The LNG chain: the project nature of the LNG business -- Upstream natural gas -- The liquefaction plant -- LNG tankers -- LNG import terminals -- The evolution of the LNG industry -- Orchestration and financing of an LNG project -- Global liquefaction projects -- The economics of an LNG project -- Upstream gas supply agreements -- Sale and purchase contracts -- LNG Tanker Contracts -- Shipping Conventions and Regulations -- Import-Related Project Siting and Regulatory Issues -- LNG Import Terminal Use Agreements -- Conclusion: the future of LNG.
Subjects	Liquefied natural gas industry.
Notes	Includes bibliographical references (p. [405]-406) and index.

Marine navigation and safety of sea transportation

LCCN	2010364373
Type of material	Book
Meeting name	International Navigational Symposium on Marine Navigation and Safety of Sea Transportation (8th: 2009: Gdynia, Poland)

Main title	Marine navigation and safety of sea transportation / editor, Adam Weintrit.
Published/Created	Boca Raton: CRC; London: Taylor & Francis [distributor], c2009.
Description	xxvi, 794 p.: ill.; 26 cm.
ISBN	9780415804790 (hbk.) 0415804795 (hbk.) 9780203869345 (ebook) 0203869346 (ebook)
LC classification	VK560 .I589 2009
Related names	Weintrit, Adam.
Contents	TransNav 2009: review of scientists and professionals meeting in the field of safety of navigation and sea transportation -- Foreword to the monograph / A. Weintrit & T. Neumann -- List of reviewers -- Message from the president of The Nautical Institute --The present and expected changes in maritime safety, security and defense functions / J. Urbański, W. Morgaś & M. Mięsikowski -- Study on ships safety control system / X.-J. Han, X.-Y. Meng & Z.-W. Wang -- Marine navigation using expert system / N. Nikitakos & G. Fikaris -- Safety of navigation and spatial planning at sea / J. Hajduk -- e-Navigation and the human element / D. Patraiko, P. Wake & A. Weintrit -- Generation of electronic nautical chart data for assessment of navigational safety in harbour and waterway design / K. Guan. C. Shi, S. Wu & T. Xu -- Study on shipping safety strategy based on accident forecast model / X.Y. Meng, Y.M. Bai & X.J. Han -- Knowledge representation in a ship's navigational decision support system / Z. Pietrzykowski & J. Uriasz -- Manoeuvring simulation on the bridge for predicting motion of

real ships and as training tool in ship handling simulators / K. Benedict, M. Kirchhoff, M. Gluch. S. Fischer & M. Baldauf -- CFD based hull hydrodynamic forces for simulation of ship manoeuvres / T. Tabaczek, T. Górnicz & J. Kulczyk -- New capabilities of the NTPRO 4000 full mission ship handling simulator in the assessment and evaluation processes at Lithuanian Maritime Academy / R. Zažeckis, I. Bartusevičienė & R. Maksimavičius -- Reconstructing a marine casualty: the effectiveness of the full-mission simulator as a casualty analysis tool / E. Doyle -- Fuzzy fast time simulation model of ship's manoeuvring / P. Zalewski -- Ship manoeuvring performance experiments using a free running model ship / N. Im & J.-H. Seo -- Simulation of load distribution along a quay during unparallel berthing manoeuvres / J. Artyszuk -- Training course for personnel involved in emergency towing operations / T.E. Berg, G. Gudmundseth & U. Klevstad -- Modernization of maritime DGPS in Poland / M. Dziewicki -- Application of 3-D velocity measurement of vessel by VI-GPS for STS lightering / Y. Yoo, E. Pedersen, K. Tatsumi, N Kouguchi & Y. Arai -- Positioning using GPS and GLONASS systems / L. Kujawa, J.B. Rogowski & K. Kopańska -- Galileo integrity concept and its applications to the maritime sector / C. Hernández, C. Catalán & M.A. Martínez -- Galileo AltBOC E5 signal characteristics for optimal tracking algorithms / E. Vejražka, P Kovář & P. Kačmařík -- The implementation of the EGNOS system to APV-I precision approach operations / A. Fellner, K. Banaszek & P

Tróminski -- GPS-based vehicle localisation / A. Janota & V. Končelík -- Effect of measurement duration on the accuracy of position determination in GPS and GPS/EGNOS systems / R. Bober, T. Szewczuk & A. Wolski -- Sustainability of motorways of the sea and fast ships / E.X. Martínez de Osés & M. Castells i Sanabra -- Applying graph theory terms to description of VTS / K. Jackowski -- Simulation-based risk analysis of maritime transit traffic in the Strait of Istanbul / B. Ozbas, I. Or, O.S. Uluscu & T Altıok -- The Marine Electronic Highway project in Straits of Malacca and Singapore: observation on the present development / M.H. Said & A.H. Saharuddin -- Availability of traffic control system based on servicing model / J. Mikulski -- Evaluation of main traffic congestion degree for restricted waters with AIS reports / Q. Hu, J. Yong, C. Shi & G. Chen -- Computer vision and ship traffic analysis: inferring maneuver patterns from the automatic identification system / K.G. Aarsæther & T. Moan -- Possible method of clearing-up the close-quarter situation of ships by means of Automatic Identification System / V.M. Bukaty & S.U. Morozova -- Development of a concept for bridge alert management / F. Motz, S. Höckel, M. Baldauf & K. Benedict -- Comparison of traditional and integrated bridge design with SAGAT / F. Motz, E. Dalinger, H. Widdel, S. Höckel & S. MacKinnon -- The problem of "infant mortality" failures of integrated navigation systems / S. Ahvenjärvi -- CRM-203 type frequency modulated continuous wave (FM CW) radar / S. Plata & R. Wawruch -- The impact of windmills on the operation of radar systems / M. Džunda, V. Humeňanský, D. Draxler, Z. Csefalvay & P. Bajusz -- 3D sonar for navigation and obstacle avoidance / I. Bowles & Z. Markowski -- The problem of magnetic compass deviation at contemporary conditions / E.M. Lushnikov -- The basic research for the new compass system using latest MEMS / G. Fukuda & S. Hayashi -- Development of decision supporting tools for determining tidal windows for deep-drafted vessels / K. Eloot, M. Vantorre, J. Richter & J. Verwilligen -- Behaviour patterns in crossing situations / J. Kemp -- Method of safe returning of the vessel to planned route after deviation from collision / M. Tsymbal & I. Urbansky -- A study of marine incidents databases in the Baltic Sea Region / A. Mullai, E. Larsson & A. Norrman -- The display mode for choosing the manoeuvre for collision avoidance / L. Vagushchenko & A. Vagushchenko -- Defining of minimally admitted head-on distance before the ships start maneuvering / V.M. Bukaty & E.N. Dimitrieva -- Collision scenario-based cognitive performance assessment for marine officers / H. Kim, H.-J. Kim & S. Hong -- The effects of causation probability on the ship collision statistics in the Gulf of Finland / M. Hänninen & P. Kujala -- An influence of the order to maintain minimum distance between successive vessels on the vessel traffic intensity in the narrow fairways / L. Kasyk -- On determination of the head-on situation under Rule 14 of COLREG-72 / V.M. Bukaty & S.U. Morozova -- Maritime communication to support safe navigation / K.E. Fjørtoft, B. Kvamstad & F. Bekkadal -- Some radiocommunication aspects of e-navigation / K. Korcz -- On-board communication challenges (LAN, SOA and wireless communication) / L. Mu & N. Garmann-Johnsen -- Towards standardized maritime language for communication at sea / B. Katarzyńska -- Novel maritime communications technologies / F. Bekkadal -- Advantages of preservation of obligatory voice communication on the VHF radio channel 16 / S. Brzóska -- The transmission of the information of the system of telecornmunicational DECT in the trans-shipping terminal / A. Kuśmińska-Fijałkowska & Z. Łukasik -- Navigational safety in SPM (single mooring point) regions / V. Paulauskas -- Identification of ship maneuvering model using extended Kalman filters / C. Shi, D. Zhao, J. Peng & C. Shen -- Estimating manoeuvres safety level of the Unity Line m/f "Polonia" ferry at the Port of Ystad / A. Kowalski -- Conceptual model of port security simulating complex (Bulgarian standpoint) / B. Mednikarov, N. Stoyanov & K. Kalinov -- Problem of stopping vessel at the waypoint for full-mission control autopilot / L. Morawski & V. Nguyen Cong -- On the control of CPP ships by steering during in-harbour ship-handling / H. Yabuki & Y. Yoshimura -- New Black Sea Terminal of port Kulevi and it navigating features / A. Gegenava, N. Varshanidze & G. Khaidarov -- Analysis of the influence of current on the manoeuvres of the turning of the ship on the ports turning-basins / J. Kornacki -- Satellite and terrestrial radionavigation systems on European inland waterways / J. Januszewski -- Electronic reporting of ships in the RIS system / A.

Lisaj -- The criterion of safety navigation assessment in sea-river shipping / W. Galor – Target tracking in RIS / A. Stateczny & W. Kazimierski -- Six in one or one in six variants: electronic navigational charts for open sea, coastal, off-shore, harbour, sea-river and inland navigation / A. Weintrit -- Data transmission in inland AIS system / P. Wołejsza. Multi-objective optimization of motor vessel route / S. Marie & E. Courteille -- Application of the 1-2-3 rule for calculations of a vessel's route using evolutionary algorithms / B. Wiśniewski, P. Medyna & J. Chomski -- Multicriteria optimisation in weather routing / J. Szłapczyńska & R. Śmierzchalski -- On the fuel saving operation for coastal merchant ships using weather routing / K. Takashima, B. Mezaoui & R. Shoji -- Solving multi-ship encounter situations by evolutionary sets of cooperating trajectories / R. Szłapczyński -- Evolutionary sets of cooperating trajectories in multi-ship encounter situations: use cases / R. Szłapczyński -- Contemporary problems of navigation nearly pole / E.M. Lushnikov -- A case study from an emergency operation in the Arctic Seas / B. Kvamstad, K.E. Fjørtojt, F. Bekkadal, A.V. Marchenko & J.L. Ervik -- Ice conditions and human factors in marine accidents at the Arctic / N. Marchenko -- Sea ice services in the Baltic Sea / M. Sztobryn -- Low sea level occurrence of the southern Baltic Sea coast / I. Stanisławczyk, B. Kowalska & M. Mykita -- Measurement system for wind and waves characteristics registration on the Silm Lake / L. Morawski, J. Pomirski, P. Sikora & R. Sokół -- Simplified method for estimating maximum ship's draught when navigating in shallow water on the south of Stolpe Bank in the aspect of the vessels with maximun dimensions and draught / G. Rutkowski & A. Królikawski -- Asymptotic theory of ship motions in regular waves under shallow water conditions / Y.L. Vorobyov & M.S. Stasenko -- Stabilization of fractional positive continuous-time linear systems in sectors of left-hand half complex plane by state-feedbacks / T. Kaczorek -- The comparison of safe control methods in marine navigation in congested waters / J. Lisowski -- A numerical study of combined natural and Marangoni convection in a square cavity / K. Cicek & A. Cihat Baytas -- An application of mathematical theory of evidence in navigation / W. Filipowicz -- The H_2 and robust H[subscript inf] regulators applied to multivariable ship steering / W. Gierusz -- Speciation of population in neuroevolutionary ship handling / M. Łącki -- Equalization of the measurements of the altitude, the azimuth and the time from observation of passages of celestial bodies / P. Bobkiewicz -- Programmatic correction of errors of measuring track processing / M. Luft, E. Szychta & R. Cioc -- Alternative for Kalman filter: two dimension self-learning filter with memory / A. Fellner, K. Banaszek & P. Tróminski. Managing and predicting maritime and off-shore risk / R.B. Duffey & J.W. Saull -- Transportation system architecture for intelligent management / J. Szpytko -- Risk analysis and human factor in prevention of CRG casualties / L. Kobyliński -- Estimation of the probability of propulsion loss by a seagoing ship based on expert opinions / A.

Brandowski & W. Frąckowiak -- Finite discrete Markov model of ship safety / L. Smolarek -- The possibility of application of algorithms indicating maximum paths in directed graphs for modeling of the evacuation process / D.H. Łozowicka -- Maritime transport development in the global scale: the main chances, threats and challenges / A.S. Grzelakawski -- Maritime safety in European concept of the internalization of external costs of transport / M. Matczak -- e-Maritime: an enabling framework for knowledge transfer and innovative information services development across the waterborne transport sector / J. Graff -- Challenges for Polish seaports' development in the light of globalisation processes in maritime transport / A. Przybyłowski -- An analysis of marine navigation and safety of sea transportation by Iranian women as officer and master mariner / H. Yousefi -- Modelling support for maritime terminals planning and operation / S. Ricci & C. Marinacci -- Turkish maritime transport policy (1960-2008) / M. Kadioglu -- The influence of organic polymer on parameters determining ability to liquefaction of mineral concentrates / M. Popek -- Application of thermal analysis and trough test for determination of the fire safety of some fertilizers containing nitrates / K. Kwiatkawska-Sienkiewicz & P. Kałucka -- Problem behaviours among children of Filipino seafarers in Iloilo City, Philippines / V.B. Jaleco, M.G. Gayo, Jr., R.L. Pador & R.A. Alimen -- Predicting emotional intelligence in maritime management: imperative, yet elusive / E.S. Potoker & J.-A. Corwin -- Officers' shortage: viewpoints from stakeholders / G. Eler, J. Calambuhay, L.

Bernas & M. Magramo -- A noble profession called seafaring: the making of an officer / M. Magramo & L. Gellada -- Officers as prostitutes: myth or reality? (a study on poaching of officers in the Philippines) / M. Magramo, G. Eler, J Calambuhay & L. Bernas -- The economical emigration aspect of East and Central European seafarers: motivation for employment in foreign fleet / V. Senčila, I. Bartusevičienė, L. Rupšienė & G. Kalvaitienė -- The role of the maritime institutions on the shortage of officers / M. Magramo, L. Bernas, J. Calambuhay & G. Eler -- Psychological features of seamen's activity in emergency situations / V.A. Bondarev & O.M Bondareva. Maritime education: putting in the right emphasis / A. Ali -- Correlation between academic performance in Auxiliary Machinery 2 subject and navigational trip among marine engineering students at Maritime University in the Philippines / R.A. Alimen, V.B. Jaleco, R.L. Pador & M.G. Gayo, Jr. -- Higher performance in maritime education through better trained lecturers / R. Hanzu-Pazara, P. Arsenie & L. Hanzu-Pazara -- Mentoring and the transfer of experiential knowledge in today's merchant fleet / A.L. Le Goubin -- Stakeholder satisfaction: research evaluation of marine engineering cadets' performance at Maritime University, Philippines / R.A. Alimen, M. Gayo, Jr. & V.B. Jaleco -- Project PRACNAV for a better on board training curricula / E. Barsan & C. Muntean -- A new tool for evaluating and training of chemical tanker crew: seafarer evaluation and training software: DEPEDES (SETS) / O. Arslan, O. Gurel & M.

Kadioglu -- MET system in Ukraine / M.V. Miyusov & D.S. Zhukov -- The Somali piracy new or old challenge for international community / D. Duda & T. Szubrycht -- The importance of the educational factor to assure the safe and security on the sea / L.C. Stan & N. Buzbuchi -- Standard for quality assurance: the case of Philippine Maritime College / A.C. Doromal -- Novelties in the development of the qualification standards for electro-technical officers under STCW convention requirements / J. Wyszkowski, J. Mindykowski & R. Wawruch -- Assessment of ISPS code compliance at ports using cognitive maps / M. Celik & Y. Ilker Topcu -- Dynamic component of ship's heeling moment due to sloshing vs. IMO IS-code recommendations / P. Krata -- The influence of the flooding damaged compartment on the metacentric height ship type 888 / W. Mironiuk -- Intelligent evaluation system of ship management / Q. Xu, X. Meng & N. Wang -- Round table panel session: GNSS and safety and security of marine navigation.

Subjects Electronics in navigation--Congresses.
Aids to navigation--Congresses.
Shipping--Safety measures--Congresses.
Shipping--Technological innovations--Congresses.
Nautical instruments--Congresses.

Notes Proceedings of the 8th International Navigational Symposium on Marine Navigation and Safety of Sea Transportation, held June 17-19, 2009 in Gdynia, Poland.
Includes bibliographical references and index.

New frontiers in engineering geology and the environment: proceedings of the International Symposium on Coastal Engineering Geology, ISCEG-Shanghai 2012

LCCN	2012942020
Type of material	Book
Meeting name	International Symposium on Coastal Engineering Geology (2012: Shanghai, China)
Main title	New frontiers in engineering geology and the environment: proceedings of the International Symposium on Coastal Engineering Geology, ISCEG-Shanghai 2012 / Yu Huang ... [et al.], editors.
Published/Created	Berlin; New York: Springer, [2012], c2013.
Description	xi, 304 p.: ill. (some col.), maps (some col.); 29 cm.
ISBN	9783642316708 (alk. paper) 3642316700 (alk. paper) 9783642316715 (ebk.)
LC classification	TA703.5 .I587 2012
Portion of title	Proceedings of the International Symposium on Coastal Engineering Geology, ISCEG-Shanghai 2012
Related names	Huang, Yu, 1973-
Contents	Co-seismic fault effects of landslides triggered by Wenchuan Ms 8.0 earthquake, China / R.Q. Huang and W.L. Li -- Large deformation analysis for costal geo-disasters using continuum and discrete modeling / A. Yashima ... [et al.] -- Seismic design of piles in liquefiable soils / S. Bhattacharya, R. Sarkar, and Y. Huang -- A new method for predicting consolidation settlement of soft ground reinforced with preloading technique / G.B. Ye, X. An, and J. Wu -- An approach to identify a linked spatial network for large mammal conservation on Yellow Sea coast / S.B. Fang ... [et al.] -- Analysis

of waterfront excavation adjacent to pile-supported wharves / Z.K. Cheng ... [et al.] -- Deformation behavior of braced excavation and its influence on adjacent piles / Y.C. Ding ... [et al.] -- Numerical analysis on deformation of braced excavation with top-down method / Y.C. Ding ... [et al.] -- Analysis of engineering geology condition for Qingdao Tongan area / D.W. Lv ... [et al.] -- Simulation of tidal flat terrain based on landform feature lines of tidal basin / Y.Y. Kang and X.R. Ding -- Test and control methods for on-site compaction of fine sand subgrade on coastal regions / J.S. Qian, H.L. Wang, and P. Wang -- Strength and permeability of biostabilized sand / E. Yee ... [et al.] -- Numerical study on the full-range consolidation characteristics of a new dredger fill foundation / F.F. Ren ... [et al.] -- Centrifuge modeling of embedment effects on eccentrically loaded shallow foundation on sand / M. Cocjin ... [et al.] -- An investigation on stress-strain relationship of on Eucalyptus saligna by pull-out method based on regression analysis / F. Chen ... [et al.] -- Detection of submerged sand bars in the Ebro Delta using ASTER images / R. Rodríguez-Martín and I. Rodríguez-Santalla -- A brief review of actual dune dynamics modeling: applicability to El Fangar dune system (Ebro Delta-Spain) / F. Barrio-Parra ... [et al.] -- Transportation of Huanghe River-discharged-suspended sediments in nearshore of Huanghe River Delta in conditions of different estuary channels / X.H. Xue, G.S. Li, and L.Y. Yuan -- 3D sediment physical model test study for PLTU 2 JATENG 1 x 660 MW Adipala, Cilacap, Indonesia / F. Gao, H.B. Chen, and Y. Luo --

Compression behaviors of marine clay for coastal reclamation in Dalian, China / M.J. Jiang, N. Zhang, and J.D. Liu -- Numerical modeling of tidal effects on groundwater in the coastal aquifer of Donghai Island / P.P. Zhou ... [et al.] -- Engineering geological assessment of the Anzali coastal region (north Iran, south Caspian coast) to sustain urban planning and development / M. Hashemi ... [et al.] -- Evaluation of engineering geological condition in Shanghai coastal area / Y.J. Shi ... [et al.] -- Study of optimum building spacing of dense high-rise building group in Shanghai based on centrifugal model test / Y.Q. Tang ... [et al.] -- THM simulation for real-scale field test / Y.L. Xiong ... [et al.] -- Cluster analysis for orientation data using DifFUZZY method / J. Wu and Z.X. Zhang -- Research progress and trend of geotechnical environmental problems in large-scale reclamation projects / F.S. Ma ... [et al.] -- Change law of vane shear strength of soft soil with depth in coastal city / P. Yang ... [et al.] -- The bearing capacity of flexible piles under combined loads in dense sand / M. Jian ... [et al.] -- Foundation design challenges at Hunter Expressway Alliance project in Australia / H.H. Zhang and B. Choi -- Analytical solutions for three-dimensional stability of coastal slope / C.Y. Han, X.H. Xia, and J.H. Wang -- NCEER method and application of an embankment slope with liquefaction problem in coastal area / G.L. Wang ... [et al.] -- Low-order lateral buckling analysis of submarine pipeline under thermal stress / L.P. Guo, R. Liu, and S.W. Yan -- DEM coupled SMAC simulation on the moving process of flow like landslide / A. Kondo, Y. Huang, and K. Maeda -- Risk assessment for coal mining under sea area / W.H. Sui and Z.M. Xu -- Potential of using remote sensing data for dike inspection / S.L. Cundill ... [et al.] -- Experimental investigation for water flowing fractured zone due to coal mining under sea area / W.X. Wang, W. Hu, and Y.K. Liang -- Geodisasteres [i.e. geo disasters] prevention and reinforcement of the sea coast in Primorsko Town in Bulgaria / K.A. Anguelov -- Potential tsunami hazard modelling of Black Sea coastline, Turkey / Ö. Acır ... [et al.] -- Alluvial risk in the coastal plain of Pesaro City, northern Marche (Italy) / U. Gori, M. Mari, and R.W. Romeo -- Analysis of the carbon monoxide released law caused by shield construction in shallow stratum in sea reclamation district / J.X. Wang ... [et al.] -- Numerical simulation of seismic responses of asymmetric-plan reinforced concrete frame with BRBs / H.J. Sun and C.X. Li -- Seismic analysis of geosynthetic-reinforced quay-wall structure / J.F. Lu, B. Ye, and J. Nagaya -- Shape effect on active earth pressure against a sheet pile wall with different displacement modes / F. Liu ... [et al.] -- "Atlant" anchor pile technology and capacity experimental results / D. Malinin and A. Malinin -- Statistical prediction of overburden failure due to coal mining under sea area / Z.M. Xu and W.H. Sui -- Numerical analysis for the stability of over-length trench wall based on a novel pre-supporting excavation method / Z.X. Zhang, C.H. Pan, and C.F. Duan -- Estimating the effects of shield tunnelling on buried pipelines based on a Kerr foundation model / H. Zhang and Z.X. Zhang --

	Depositional history and geotechnical properties of Shanghai clays / C.J. Wu ... [et al.] -- On the using of deformation rates as stability control parameters of PVD-treated soft ground / Y.B. Gao -- Application of improved vacuum preloading method in Tianjin Lingang industrial zone / C. Le ... [et al.] -- Observations on the new trace of Quaternary glacial epoch in the Yangtze Three Gorges area / Y.J. Li -- Strengths of unsaturated silty clay used as garden hill fill in Shanghai / Y.B. Gao, X.N. Ge, and J. Song.
Subjects	Engineering geology--Congresses. Coastal engineering--Congresses. Environmental engineering--Congresses.
Notes	International Symposium on Coastal Engineering Geology held in Shanghai, China, Sept. 20-21, 2012. Cf. pref. Includes bibliographical references and indexes.
Series	Springer geology Springer geology.

Paleoseismology

LCCN	2009013122
Type of material	Book
Main title	Paleoseismology / edited by James P. McCalpin.
Edition	2nd ed.
Published/Created	Burlington, MA: Academic Press, c2009.
Description	xvi, 613 p., [32] p. of plates: ill. (some col.), maps (some col.); 24 cm.
Links	Publisher description http://www.loc.gov/catdir/enhancements/fy1011/2009013122-d.html
ISBN	9780123735768 0123735769
LC classification	QE539.2.P34 P35 2009

Related names	McCalpin, James.
Contents	Introduction to paleoseismology -- Field techniques in paleoseismology - terrestrial environments -- Sub-aqueous paleoseismology -- Paleoseismology in extensional tectonic environments -- Paleoseismology of volcanic environments -- Paleoseismology of compressional tectonic environments -- Paleoseismology of strike-slip tectonic environments -- Using liquefaction-induced and other soft-sediment features for paleoseismic analysis -- Using landslides for paleoseismic analysis.
Subjects	Paleoseismology.
Notes	Includes bibliographical references and index.
Series	International geophysics series; v. 95 International geophysics series; v. 95.

Pathogenesis of human pulmonary tuberculosis: insights from the rabbit model

LCCN	2006021123
Type of material	Book
Personal name	Dannenberg, Arthur M.
Main title	Pathogenesis of human pulmonary tuberculosis: insights from the rabbit model / Arthur M. Dannenberg.
Published/Created	Washington, D.C.: ASM Press, c2006.
Description	xiv, 453 p.: ill.; 27 cm.
Links	Table of contents only http://www.loc.gov/catdir/toc/ecip0616/2006021123.html
ISBN	1555813739 9781555813734 9781555813734

LC classification RC311 .D362 2006
Related names American Society for Microbiology.
Contents Stages in the pathogenesis of human and rabbit tuberculosis -- Types of human pulmonary tuberculosis -- Liquefaction of caseous foci and cavity formation -- Delayed-type hypersensitivity, cell-mediated immunity, and antibodies in tuberculosis -- Macrophages and other cells in tuberculous lesions -- Structural components of tuberculous lesions -- Microvascular density in tuberculous lesions -- Early pulmonary lesions in rabbits -- Macrophage turnover, division, and activation in tuberculous lesions -- Lurie's pulmonary Tubercle-count method -- Natural airborne infection -- Response of rabbits to inhaled tubercle bacilli including BCG-- Characteristics of resistance and susceptibility to tuberculosis in Lurie's inbred rabbits -- Comparisons of tuberculosis in rabbits, mice, and guinea pigs -- Effects of cortisone and adrenocorticotropic hormone on tuberculosis -- Effects of estrogen, chorionic gonadotropin, and thyroid hormone on tuberculosis -- Effects of whole-body X-irradiation on tuberculosis -- Cytokine production in primary BCG lesions -- Cytokine production in reinfection bcg lesions and in tuberculin reactions -- Vascular adhesion molecules in tuberculous lesions -- Principles and guidelines for developing better tuberculosis vaccines -- Characteristics of rabbit BCG lesions and efficacies of BCG and mycobacterium microti vaccines.
Subjects Lurie, Max B. (Max Bernard), 1893-1966.
Tuberculosis--Animal models.
Tuberculosis--Pathophysiology.

	Tuberculosis, Pulmonary--physiopathology.
	Models, Animal.
	Mycobacterium bovis--pathogenicity.
	Rabbits.
	Tuberculosis, Pulmonary--microbiology.
Notes	Includes bibliographical references and index.

Practices and procedures for site-specific evaluations of earthquake ground motions

LCCN	2012932145
Type of material	Book
Personal name	Matasovic, Neven, 1959-
Main title	Practices and procedures for site-specific evaluations of earthquake ground motions / consultants, Neven Matasovic and Youssef Hashash.
Published/Created	Washington, D.C.: Transportation Research Board, 2012.
Description	78 p.: ill., forms; 28 cm.
Links	http://onlinepubs.trb.org/onlinepubs/nchrp/nchrp_syn_428.pdf
ISBN	9780309223553 (pbk.) 0309223555 (pbk.)
LC classification	TE208 .M38 2012 QE539.2.S34 M38 2012
Portion of title	Site-specific evaluations of earthquake ground motions
Related names	Hashash, Youssef M. A. National Research Council (U.S.). Transportation Research Board. National Cooperative Highway Research Program. American Association of State Highway and Transportation Officials. United States. Federal Highway Administration.

Summary The current AASHTO specifications for seismic design mandate site-specific evaluation of the earthquake design ground motion (i.e., the acceleration response spectrum) for ground conditions termed Site Class F. In the AASHTO specifications, Site Class F soils include soft clay sites. These AASHTO specifications also allow discretionary site-specific analyses for other ground conditions and a reduction in mapped design ground motions of as much as 33% if justified by a site-specific ground motion analysis. Some state departments of transportation (DOTs) are taking advantage of this site response reduction provision, particularly in cases where pore pressure generation could lead to soil liquefaction- Source other than Library of Congress.

Contents Introduction -- Current state of evaluation of site effects on ground motions -- Approach to survey of current practice -- Survey responses and relevant literature -- Conclusions and suggestions for further research

Subjects Soils--United States--Testing.
Pavements--United States--Subgrades--Testing.
Shear strength of soils--Measurement.
Earthquake hazard analysis.
Earthquake resistant design.
Soils--Vibration.
Seismology--United States.
Earthquakes--United States.
Design methods.
Geological activities and processes.
Ground movements.
Planning stages.
Seismic design.

	Seismicity.
Notes	At head of title: National Cooperative Highway Research Program. "A synthesis of highway practice." Includes bibliographical references (p. 27-38). Research sponsored by the American Association of State Highway and Transportation Officials in cooperation with the Federal Highway Administration NCHRP project 20-05, topic 42-03
Additional formats	Also available online. Online version: Matasovic, Neven, 1959- Practices and procedures for site-specific evaluations of earthquake ground motions. Washington, D.C.: Transportation Research Board, 2012 (OCoLC)794445419
Series	NCHRP synthesis, 0547-5570; 428 Synthesis of highway practice; 428.

Producing liquid fuels from coal: prospects and policy issues

LCCN	2008050794
Type of material	Book
Personal name	Bartis, James T., 1945-
Main title	Producing liquid fuels from coal: prospects and policy issues / James T. Bartis, Frank Camm, David S. Ortiz.
Published/Created	Santa Monica, CA: Rand Corp., 2008.
Description	xxx, 167 p.: ill. (some col.); 26 cm.
Links	Table of contents only http://www.loc.gov/catdir/toc/fy0906/2008050794.html
ISBN	9780833045119 (pbk.: alk. paper) 0833045113 (pbk.: alk paper)
LC classification	TP352 .B37 2008
Related names	Camm, Frank A., 1949- Ortiz, David (David Santana)

Contents	Introduction -- The coal resource base -- Coal-to-liquids technologies -- Other unconventional fuels -- Benefits of coal-to-liquids development -- Critical policy issues for coal-to-liquids development -- Designing incentives to encourage private investment -- Moving forward with a coal-to-liquids development effort -- Appendixes: A. Cost-estimation methodology and assumptions -- B. Greenhouse-gas emissions: supporting analysis -- C. A model of the global liquid-fuel market.
Subjects	Coal liquefaction. Coal liquefaction--Government policy--United States.
Notes	"Rand Project Air Force and Infrastructure, Safety, and Environment." Includes bibliographical references (p. 155-167).

Shear wave propagation and liquefaction in layered [sic] soils = [Tabakalı zeminlerde kayma dalgası yayılması ve sıvılaşma]

LCCN	2002471325
Type of material	Book
Personal name	Tezcan, Semih S.
Main title	Shear wave propagation and liquefaction in layered [sic] soils [Tabakalı zeminlerde kayma dalgası yayılması ve sıvılaşma] / Semih S. Tezcan, Leon Teri.
Published/Created	Maslak, İstanbul: Turkish Earthquake Foundation, 2001.
Description	150 p.: ill., maps; 30 cm.
LC classification	TA710 .T46 2001
Variant title	Shear wave propagation and liquefaction in layered soils

Parallel title	Tabakalı zeminlerde kayma dalgası yayılması ve sıvılaşma [Shear wave propagation and liquefaction on layered substrates]
Related names	Teri, Leon.
	Türkiye Deprem Vakfı.
Subjects	Soil liquefaction.
	Shear waves.
	Earthquake engineering.
Notes	At head of title: Turkish Earthquake Foundation
	Includes bibliographical references (p. [106]-110).
	In English with a summary also in Turkish.
Series	Kitap / TDV; KT 016-64
	Kitap (Türkiye Deprem Vakfı); KT 016-64.

Small-scale gas to liquid fuel synthesis

LCCN	2015410238
Type of material	Book
Main title	Small-scale gas to liquid fuel synthesis / edited by Nick Kanellopoulos.
Published/Produced	Boca Raton: CRC Press, [2015]
Description	xvii, 570 pages: illustrations (some color); 24 cm
ISBN	9781466599383
	1466599383
LC classification	TP761.L5 S58 2015
Related names	Kanellopoulos, N. K. (Nick K.), editor.
Subjects	Liquefied natural gas.
	Gases--Liquefaction.
	Liquefied natural gas industry.
	Business & Economics / Industries / General
	Liquefied natural gas.
	Liquefied natural gas industry.
Form/Genre	Electronic books.
Notes	Includes bibliographical references and index.

Soil dynamics and liquefaction 2000: proceedings of sessions of Geo-Denver 2000: August 5-8, 2000, Denver, Colorado

LCCN	00042133
Type of material	Book
Meeting name	Geo-Denver 2000 (2000: Denver, Colo.)
Main title	Soil dynamics and liquefaction 2000: proceedings of sessions of Geo-Denver 2000: August 5-8, 2000, Denver, Colorado / sponsored by The Geo-Institute of the American Society of Civil Engineers; edited by Ronald Y.S. Pak, Jerry Yamamura.
Published/Created	Reston, Va.: American Society of Civil Engineers, c2000.
Description	vii, 205 p.: ill.; 22 cm.
ISBN	0784405204
LC classification	TA711.A1 G46 2000a
Related names	Pak, Ronald Y. S. Yamamura, Jerry. American Society of Civil Engineers. Geo-Institute.
Subjects	Soil dynamics--Congresses. Soil liquefaction--Congresses. Soil structure interaction--Congresses.
Notes	Includes bibliographical references and indexes.
Series	Geotechnical special publication; no. 107

Soil liquefaction: a critical state approach

LCCN	2006002052
Type of material	Book
Personal name	Jefferies, Mike, 1952-
Main title	Soil liquefaction: a critical state approach / Mike Jefferies & Ken Been.
Published/Created	London; New York: Taylor & Francis, 2006.
Description	xxxi, 479 p.: ill.; 26 cm.

Links	Table of contents only http://www.loc.gov/catdir/toc/ecip067/2006002052.html Publisher description http://www.loc.gov/catdir/enhancements/fy0651/2006002052-d.html
ISBN	0419161708 (hbk.) 9780419161707 (hbk.) 020330196X (ebk.) 9780203301968 (ebk.)
LC classification	TA710 .J365 2006
Related names	Been, Ken.
Subjects	Soil mechanics. Soil liquefaction.
Notes	Includes bibliographical references (p. [459]-475) and index.

Soil liquefaction: a critical state approach

LCCN	2015490976
Type of material	Book
Personal name	Jefferies, Mike, 1952- author.
Main title	Soil liquefaction: a critical state approach / Mike Jeffries, Golder Associates (UK), Canada, Ken Been, Golder Associates, Houston, USA.
Edition	Second edition.
Published/Produced	Boca Raton: CRC Press, Taylor & Francis Group, [2016]
Description	xxi, 690 pages: illustrations, charts; 27 cm
ISBN	1482213680 9781482213683
LC classification	TA710 .J365 2016
Related names	Been, Ken, author.
Summary	"Soil Liquefaction has become widely cited. It is built on the principle that liquefaction can, and must, be understood from mechanics. This second edition is developed from this premise in three

respects: with the inclusion of silts and sandy silts commonly encountered as mine tailings, by an extensive treatment of cyclic mobility and the cyclic simple shear test, and through coverage from the "element" scale seen in laboratory testing to the evaluation of "boundary value problems" of civil and mining engineering. As a mechanics-based approach is necessarily numerical, detailed derivations are provided for downloadable open-code software (in both Excel/VBA and C++) including code verifications and validations. The "how-to-use" aspects have been expanded as a result of many conversations with other engineers, and these now cover the derivation of soil properties from laboratory testing through to assessing the in situ state by processing the results of cone penetration testing." -- from publisher.

Contents 1. Introduction -- 2. Dilatancy and the state parameter -- 3. Constitutive modelling for liquefaction -- 4. Determining state parameter in-situ -- 5. Soil variability and characteristic states -- 6. Static liquefaction and post-liquefaction strength -- 7. Cyclic stress-induced liquefaction (cyclic mobility and softening) -- 8. Finite element modelling of soil liquefaction -- 9. Practical implementation of critical state approach -- 10. Concluding remarks.

Subjects Soil liquefaction.
Soil mechanics.
Soil liquefaction.
Soil mechanics.

Notes Includes bibliographical references (pages 659-676) and index.

Series	Applied geotechnics series Applied geotechnics.

Super simple earthquake projects: science activities for future seismologists

LCCN	2017946431
Type of material	Book
Personal name	Alkire, Jessie, author.
Main title	Super simple earthquake projects: science activities for future seismologists / Jessie Alkire; consulting editor, Diane Craig, M.A./reading specialist.
Published/Produced	Minneapolis, Minnesota: Super Sandcastle, an imprint of Abdo Publishing, [2018] ©2018
Description	32 pages: color illustrations; 20 x 24 cm.
ISBN	9781532112362 (hbk.) 153211236X (hbk.)
LC classification	QE521.3 .A44 2018 J551.22ALKIRE
Summary	Read all about seismology in Super Simple Earthquake Projects. Kids will learn about what causes earthquakes and how they affect Earth. Discover how scientists study and help people prepare for them. Then, build an earthquake-proof structure, create a tsunami in a bottle, and more. Each project has color photos and easy?to?follow instructions. -- Amazon.
Contents	What is an earthquake? -- What causes earthquakes? -- How scientists study earthquakes -- Materials -- Graham cracker tectonic plates -- Seismic wave box -- Seismograph -- Earthquake-proof structures -- Liquefaction experiment -- Tsunami in a bottle.

Subjects	Earthquakes--Experiments--Juvenile literature.
	Science projects--Juvenile literature.
	Earthquakes--Experiments.
	Science projects.
	Science projects--Juvenile literature.
	Earth sciences--Experiments--Juvenile literature.
	Science projects.
Form/Genre	Juvenile works.
Series	Super simple earth investigations
	Alkire, Jessie. Super simple earth investigations.

Synthetic liquids production and refining

LCCN	2011043254
Type of material	Book
Main title	Synthetic liquids production and refining / Arno de Klerk, David L. King, editor[s]; sponsored by the ACS Division of Fuel Chemistry.
Published/Created	Washington, DC: American Chemical Society, c2011.
Description	x, 326 p.: ill.; 24 cm.
ISBN	9780841226814
LC classification	TP360 .S975 2011
Related names	Klerk, Arno de.
	King, David L.
	American Chemical Society. Division of Fuel Chemistry.
Subjects	Synthetic fuels.
	Gases--Liquefaction.
	Coal--Liquefaction.
	Synthetic fuels--Refining.
Notes	Includes bibliographical references and indexes.
Series	ACS symposium series; 1084

The benefits and challenges of producing liquid fuel from coal: the role for federal research: hearing before the Subcommittee on Energy and Environment, Committee on Science and Technology, House of Representatives, One Hundred Tenth Congress, first session, September 5, 2007.

LCCN	2008377256
Type of material	Book
Corporate name	United States. Congress. House. Committee on Science and Technology (2007). Subcommittee on Energy and Environment.
Main title	The benefits and challenges of producing liquid fuel from coal: the role for federal research: hearing before the Subcommittee on Energy and Environment, Committee on Science and Technology, House of Representatives, One Hundred Tenth Congress, first session, September 5, 2007.
Published/Created	Washington, D.C.: U.S. G.P.O.: For sale by the Supt. of Docs., U.S. G.P.O., 2008.
Description	iv, 100 p.: ill.; 24 cm.
Links	http://purl.access.gpo.gov/GPO/LPS94438
ISBN	9780160807497 0160807492
LC classification	KF27 .S3935 2007j
Subjects	Coal liquefaction--Research--United States. Coal liquefaction--Economic aspects--United States. Coal liquefaction--Environmental aspects--United States. Fischer-Tropsch process. Synthetic fuels--Technological innovations--Research--United States.
Notes	Distributed to some depository libraries in microfiche.

	Shipping list no. unavailable.
	"Serial no. 110-51."
	Includes bibliographical references.
Reproduction no./Source	552-070-37353-5 GPO
Additional formats	Also available via Internet from the GPO Access web site. Address as of 5/20/08: http://frwebgate.access.gpo.gov/cgi-bin/getdoc.cgi?dbname=110_house_hearings&docid=f:37639.pdf ; current access is available via PURL.

The mechanics of scour in the marine environment

LCCN	2005299355
Type of material	Book
Personal name	Sumer, B. Mutlu.
Main title	The mechanics of scour in the marine environment / B. Mutlu Sumer, Jørgen Fredsøe.
Published/Created	River Edge, N.J.: World Scientific, c2002.
Description	xiii, 536 p.: ill.; 23 cm.
ISBN	9810249306
LC classification	TC175.2 .S85 2002
Related names	Fredsøe, Jørgen.
Summary	Publisher description: Treats the subject of local scour around different kinds of marine structures, exposed to waves and/or currents. Describes in detail the hydrodynamic processes causing the erosion.
Contents	Introduction and basic concepts -- Scour below pipelines -- Scour around a single slender pile -- Scour around a group of slender piles -- Examples of more complex configurations -- Scour around large piles -- Scour around breakwaters -- Scour at seawalls -- Ship-propeller scour -- Impact of

	liquefaction -- Appendix A: Small amplitude, linear waves.
Subjects	Scour (Hydraulic engineering) Sedimentation and deposition. Offshore structures--Foundations.
Notes	Includes bibliographical references and indexes.
Series	Advanced series on ocean engineering; v. 17

Thermochemical conversion of biomass for the production of energy and chemicals

LCCN	2015957826
Type of material	Book
Personal name	Dufour, Anthony, author.
Main title	Thermochemical conversion of biomass for the production of energy and chemicals / Anthony Dufour.
Published/Produced	Hoboken, New Jersey: Wiley, 2016.
Description	vii, 158 pages: illustrations; 25 cm
ISBN	9781119137702 (electronic bk.) 1119137705 (electronic bk.) 1848218230 (Cloth) 9781848218239 (Cloth) 1119137713 (electronic bk.) 9781119137719 (electronic bk.) 1119137691 9781119137696
LC classification	TP360 .D84 2016
Summary	This book highlights the processes of biomass thermochemical conversion, covering topics from combustion and gasification, to pyrolysis and liquefaction.Heat, power, biofuels and green chemicals can all be produced by these thermochemical processes. The different scales of investigation are presented: from the bioenergy

	chains, to the reactors and molecular mechanisms.The author uses current research and data to present bioenergy chains from forest to final use, including the biomass supply chains, as well as the life cycle assessment of different process chains.Biomass conversion reactors are also presented, detailing their technologies for combustion, gasification and syngas up-grading systems, pyrolysis and bio-oil upgrading.The physical-chemical mechanisms occurring in all these reactors are presented highlighting the main pathways for gas, char and bio-oil formation from biomass.This book offers an overview of biomass valorization for students, engineers or developers in chemistry, chemical, environmental or mechanical engineering.
Contents	Bioenergy chain and process scales -- Reactor scale -- Particle scale and mesoscale -- Molecular scale
Subjects	Biomass. Biomass chemicals. Biomass. Biomass chemicals.
Notes	Includes bibliographical references (pages 139-155) and index.
Series	Focus energy series Focus energy series.

Unconventional fossil-based fuels: economic and environmental trade-offs

LCCN	2008036873
Type of material	Book

Main title	Unconventional fossil-based fuels: economic and environmental trade-offs / Michael Toman ... [et al.].
Published/Created	Santa Monica, CA: RAND Corporation, c2008.
Description	xxiii, 72 p.: ill. (chiefly col.); 28 cm.
Links	Table of contents only http://www.loc.gov/catdir/toc/fy0903/2008036873.html
ISBN	9780833045645 (pbk.: alk. paper)
LC classification	TN871 .U49 2008
Related names	Toman, Michael A. Rand Corporation.
Subjects	Petroleum engineering. Heavy oil. Oil sands. Coal liquefaction.
Notes	Includes bibliographical references (p. 65-72).

Using Aspen plus in thermodynamics instruction: a step-by-step guide

LCCN	2014029629
Type of material	Book
Personal name	Sandler, Stanley I., 1940- author.
Main title	Using Aspen plus in thermodynamics instruction: a step-by-step guide / Stanley I. Sandler, Department of Chemical and Biomolecular Engineering, University of Delaware, Newark, DE
Published/Produced	Hoboken, New Jersey: Wiley, [2015] ©2015
Description	xi, 343 pages: illustrations; 26 cm
ISBN	9781118996911 (paperback) 1118996917 (paperback)
LC classification	QD504 .S26 2015
Summary	"This book provides the reader with a self-study, step-by-step guide to doing thermodynamic

calculations in Aspen Plus (Version 8.x). This is done by providing actual screen shots of the Aspen Plus interface to solve example problems of various types, including vapor-liquid, liquid-liquid, vapor-liquid-liquid and chemical reaction equilibria, and simple applications to liquefaction, distillation and liquid-liquid extraction"-- Provided by publisher.

Contents Getting started with Aspen Plus -- Two simpe simulations -- Pure component property analysis -- The NIST thermodata engine (TDE) -- Vapor-liquid equilibrium calculations using activity coefficient models -- Vapor-liquid equilibrium calculations using an equations of state -- Regression of liquid-liquid equilibrium (LLE) data and vapor-liquid-liquid equilibrium (VLLE) and predictions -- The property methods assistant and property estimation -- Chemical reaction equilibrium in Aspen Plus -- Shortcut distillation calculations -- A rigorous distillation calculation: RadFrac -- Liquid-liquid extraction -- Sensitivity analysis: a tool for repetitive calculations -- Electrolyte solutions.

Subjects Aspen plus.
Thermodynamics--Data processing.
Science / Mechanics / Dynamics / Thermodynamics.

Notes Includes index.

Additional formats Online version: Sandler, Stanley I., 1940- author. Using Aspen plus in thermodynamics instructions Hoboken, New Jersey: John Wiley and Sons, Inc., [2015] 9781118997000 (DLC) 2014030304

RELATED NOVA PUBLICATIONS

MODELING OF SEISMIC LIQUEFACTION DATA USING EXTREME LEARNING MACHINE*

***Pijush Samui*[1,†], *Sanjiban Sekhar Roy*[2,‡], *Pradeep Kurup*[3,¶] and *Yıldırım Dalkiliç*[4,#]**

[1]Associate Professor, Department of Civil Engineering, NIT Patna, Bihar, India

[2]Assistant Professor, SCOPE, VIT University, Vellore, India

[3]Professor, Department of Civil and Environmental Engineering, University of Massachusetts Lowell, Lowell, MA, US

[4]Assistant Professor, Faculty of Engineering, Civil Engineering Department, Erzincan University, Turkey

---

* The full version of this chapter can be found in *Earthquakes: Monitoring Technology, Disaster Management and Impact Assessment*, edited by Wayne Coleman, published by Nova Science Publishers, Inc, New York, 2016.

† Corresponding Author Email: pijush.phd@gmail.com.

‡ Corresponding Author Email: sanjibanroy09@gmail.com.

¶ Corresponding Author Email: Pradeep_Kurup@uml.edu.

Corresponding Author Email: yildirim.dalkilic@gmail.com.

Seismic liquefaction is a major concern for any earthquake prone area. This article employs a new technique based on Extreme Learning Machine (ELM) for determination of liquefaction susceptibility of soil based on Standard Penetration Test (SPT) and Cone Penetration Test (CPT) from the Chi-Chi earthquake. ELM is the modified version of Single-hidden Layer Feed forward Neural Network. Different input combinations have been tried to get best performance. A comparative study has been carried out between the SPT and CPT based models. CPT based models give better performance than the SPT based model. The developed ELM model shows that Cone Resistance (q_c) and Peck Ground Acceleration (PGA) are sufficient parameters for determination of liquefaction susceptibility of soil.

HYDROTHERMAL LIQUEFACTION OF BIOMASS FOR BIOFUEL*

***Florin G. Barla and Sandeep Kumar*†**
Department of Civil and Environmental Engineering,
Old Dominion University, Norfolk, VA, USA

In recent years, significant research interests have been placed on the conversion of non-food biomass resources such as lignocellulosic biomass (terrestrial) and algae (aquatic) to biofuels. However, high moisture content of fresh biomass or algae is one of the major challenges in developing logistics and downstream processing for an economical design of bio-refineries. Hydrothermal liquefaction (HTL) of biomass has a number of advantages including high throughputs, high energy and separation

* The full version of this chapter can be found in *Valorization of Lignocellulosic Biomass in a Biorefinery: From Logistics to Environmental and Performance Impact*, edited by Rajeev Kumar, Seema Singh and Venkatesh Balan, published by Nova Science Publishers, Inc, New York, 2016.

† Corresponding Author Email: skumar@odu.edu.

efficiencies, and feedstock flexibility. The production of liquid fuels from biomass can be achieved by HTL to an intermediate product (bio-oil or biocrude) followed by catalytic upgradation of the bio-oil/biocrude. Generally, HTL is conducted in seconds or minutes of residence time in the temperature range of 250-350°C under subcritical water conditions in a batch/flow-through/ continuous flow reactor. In this chapter, we review the recent developments of HTL of lignocellulosic biomass and microalgae along with the techno-economic analysis (TEA) and life cycle assessments (LCA).

Biomass to Bio-Oil by Liquefaction*

Huamin Wang[1,†] and Yong Wang[1,2]
[1]Pacific Northwest National Labortatory, Richland, WA, US
[2]Voiland School of Chemical Engineering and Bioengineering, Washington State University, Pullman, WA, US

Significant efforts have been devoted to develop processes for the conversion of biomass, an abundant and sustainable source of energy, to liquid fuels and chemicals in order to replace diminishing fossil fuels and mitigate global warming. Thermochemical and biochemical methods have attracted the most attention. Among the thermochemical processes, pyrolysis and liquefaction are the two major technologies for the `direct conversion of biomass to produce a liquid product, often called bio-oil. This chapter focuses on liquefaction, a medium-temperature and high-pressure thermochemical process for the conversion of biomass to bio-oil. Water has

* The full version of this chapter can be found in *Biomass Processing, Conversion and Biorefinery*, edited by Bo Zhang and Yong Wang, published by Nova Science Publishers, Inc, New York, 2013.

† Corresponding Author Phone 509-371-6705. Fax 509-372-4732. E-mail: huamin.wang@pnnl.gov.

been most commonly used as a solvent and the process is known as hydrothermal liquefaction (HTL). Fundamentals of the HTL process, key factors determining HTL behavior, the role of catalysts in HTL, properties of the produced bio-oil, and the current status of the technology are summarized. The liquefaction of biomass using organic solvents, a process called solvolysis, is also discussed. A wide range of biomass feedstocks have been tested for liquefaction including wood, crop residues, algae, food processing waste, and animal manure.

HYDROTHERMAL LIQUEFACTION- A CORE TECHNOLOGY FOR THE PRODUCTION OF ENVIRONMENT-ENHANCING ENERGY (E^2-ENERGY) *

Zhichao Wang †

Hydrothermal liquefaction (HTL), also called hydro-pyrolysis, hydrothermolysis, hydrothermal pyrolysis, hydroliquefaction, hydrothermal upgrading (HTU), or hydrothermal processing (HTP) by some researchers, is one of the thermochemical conversion processes. In this process, organic materials are converted into complicated products (of which a viscous and dark water-immiscible liquid is usually the major intended product known as bio-oil) in pressurized sub- or supercritical water.

* The full version of this chapter can be found in *Oil: Production, Consumption and Environmental Impact*edited by Shuangning Xiu, published by Nova Science Publishers, Inc, New York, 2012.

† Corresponding Author Email: zwang42@illinois.edu, Phone:1-217-493-2297.

INDEX

A

accident, vii, viii, 45, 46, 47, 48, 49, 50, 51, 52, 60, 69, 70, 71, 168

B

biodiesel, 113, 151
bioenergy, 197
biofuel, 113, 151
biogas, 139

C

C3MR, v, ix, 79, 80, 81, 82, 84, 85, 89, 90, 91, 92, 93, 95, 96, 97, 98, 99, 100
carbon, 40, 41, 42, 80, 181
carbon dioxide (CO2), 39, 40, 41, 42, 43
coal, 41, 117, 123, 124, 125, 132, 149, 159, 181, 187, 188, 194, 195
coal liquefaction, 159
cryogenic process, 2, 39

D

defence-in-depth, 49, 50, 70
dual pressure Claude, 6, 7

E

earthquakes, viii, 45, 46, 47, 48, 50, 63, 64, 69, 70, 72, 103, 107, 108, 109, 127, 155, 193
economic analysis, ix, 80, 84, 88, 92, 203
energy, vii, x, 1, 2, 3, 4, 11, 23, 26, 39, 40, 41, 42, 43, 60, 62, 80, 83, 84, 85, 86, 88, 91, 94, 95, 96, 98, 99, 100, 101, 106, 107, 114, 122, 123, 124, 125, 130, 131, 140, 143, 150, 151, 152, 159, 166, 197, 198, 202, 203
energy consumption, x, 3, 11, 26, 80, 84, 94, 96, 99, 150
energy density, vii, 1, 2
energy efficiency, 123, 124, 125
Energy Policy Act, 123, 124, 125, 158
energy recovery, 41, 43
exergoeconomic, ix, 29, 41, 43, 80, 88, 89, 91, 92, 93, 95, 96, 98, 100
exergoeconomic analysis, ix, 80, 88, 91, 92, 95, 96, 98, 100
exergy, v, vii, viii, ix, 2, 3, 16, 20, 21, 22, 23, 29, 31, 33, 39, 40, 41, 42, 43, 79, 80,

83, 84, 85, 86, 87, 88, 89, 91, 92, 93, 94, 95, 96, 97, 98, 99, 100
exergy analysis, ix, 2, 3, 33, 41, 42, 80, 86, 88, 91, 93, 98, 99
exergy-based method, vii, ix, 79, 80, 96, 101

G

gasification, 41, 117, 144, 151, 159, 197
genetic algorithm, ix, 80, 85, 94
Great Britain, 36
greenhouse, 131
grids, 2
groundwater, 135, 163, 180
growth, ix, 79, 80, 166
guidance, 56, 62
guidelines, 184
Guyana, 15

H

hazard, v, vii, viii, 45, 46, 47, 48, 49, 50, 51, 53, 54, 55, 56, 57, 63, 64, 68, 69, 70, 72, 73, 74, 75, 77, 112, 127, 128, 129, 152, 153, 157, 158, 162, 163, 164, 181, 186
helium, 8, 17, 18, 19, 20, 21, 27, 104
Helium-precooled Claude, 6
hydrocarbons, 145
hydroelectric power, 136
hydrogen, vii, 1, 2, 3, 4, 5, 6, 7, 8, 11, 12, 16, 17, 18, 19, 20, 23, 26, 27, 28, 31, 37, 39, 41, 42, 159
hydrogen gas, 18
hydrogen liquefaction, vii, viii, 2, 3, 4, 5, 8, 11, 12, 13, 14, 16, 17, 18, 19, 20, 21, 22, 23, 26, 27, 28, 39
hydrogenation, 133
hydrothermal process, 204
hypersensitivity, 184

K

Kapitza, 6, 7, 32

L

liquefied natural gas, vii, ix, 40, 41, 42, 43, 80, 91
liquid fuels, 142, 144, 187, 203
liquid hydrogen, vii, 1, 2, 39
liquids, 41, 42, 87, 117, 123, 124, 125, 147, 152, 188, 194
LNG, ix, 18, 33, 41, 42, 43, 79, 80, 81, 82, 83, 84, 85, 89, 90, 91, 97, 98, 99, 100, 101, 116, 146, 148, 162, 166, 167

N

natural gas, vii, 9, 41, 42, 43, 80, 82, 84, 89, 91, 99, 100, 116, 134, 141, 146, 148, 152, 161, 162, 166, 167, 189
natural gas pipeline, 116, 148, 161, 162
nuclear power plant, v, vii, viii, 45, 46, 49, 59, 70, 71, 72, 73, 74, 75, 76
Nuclear Regulatory Commission, 73, 157
nuclear spins, 3, 4

O

oil, 39, 133, 144, 151, 159, 198, 199, 203, 204
optimization, v, vii, ix, 30, 31, 34, 38, 39, 40, 41, 42, 43, 79, 80, 84, 85, 94, 95, 96, 97, 98, 99, 100, 101, 173
optimum cost, ix, 80

P

petroleum, 29, 33, 43, 111, 119, 133, 144, 150, 152, 199

plants, vii, viii, ix, 2, 3, 11, 14, 27, 46, 47, 48, 51, 70, 72, 80, 81, 82, 83, 99, 135, 150
plants of hydrogen liquefaction, 11
polymer, 144, 145, 175
power generation, 81, 93, 135
power plants, ix, 39, 46, 49, 59, 71, 73, 76, 91, 159
pre-cooled dual pressure Claude, 6, 13, 27
Pre-cooled dual- pressure Linde–Hampson, 6
Pre-cooled Linde–Hampson, 6
pre-cooled mixed refrigerant, ix, 80, 100
probabilistic safety analysis, viii, 46, 49, 52, 54, 64, 70, 71, 73
probability, viii, 45, 48, 51, 52, 53, 54, 55, 56, 61, 63, 64, 65, 66, 70, 171
probability distribution, 53, 56, 61, 64
propane, 81, 82, 85, 89, 91, 98, 100
propane pre-cooled mixed refrigerant liquefaction cycle, 81

R

renewable energy, viii, 2, 23, 26, 28, 130
renewable energy technologies, 26
renewable fuel, 150, 159

S

safety, v, vii, viii, 17, 45, 46, 47, 48, 49, 50, 51, 52, 53, 54, 56, 59, 60, 61, 64, 65, 66, 67, 68, 69, 70, 71, 72, 73, 74, 75, 105, 135, 141, 167, 168, 177, 188
seismic margin analysis, 52
severe accident analysis, ix, 46, 51, 69, 71
Simple Claude, 6, 13, 27
simulation, 58, 84, 90, 91, 94, 95, 118, 168, 180, 200

T

thermal analysis, 175
thermodynamic calculations, 199
thermodynamic equilibrium, 86
thermodynamics, 88, 131, 199, 200
transportation, vii, 1, 2, 107, 133, 144, 147, 167, 168, 186

W

waste, 39, 42, 113, 127, 131, 143, 144, 151, 204
waste disposal, 140
waste heat, 42

Alternative Fuels and Advanced Technology Vehicles: Incentives and Considerations

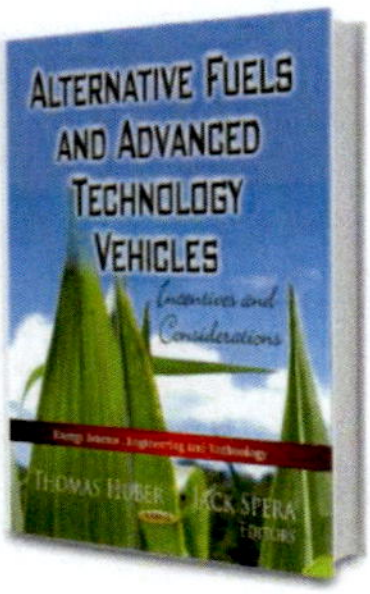

Editors: Thomas Huber and Jack Spera

Series: Energy Science, Engineering and Technology

Book Description: This book examines the current array of incentives, which do not reflect a single, comprehensive strategy, but rather an aggregative approach to a range of discreet public policy issues, including improving environmental quality, expanding domestic manufacturing, and promoting agriculture and rural developments.

Hardcover ISBN: 978-1-62257-556-5
Retail Price: $110

Transportation Energy Futures: Underexplored Topics in Vehicle Technologies and Alternative Fuel Use

Editors: Stewart Fischer and Tina Karahalios

Series: Energy Science, Engineering and Technology

Book Description: The Transportation Energy Futures Study examines technological, fuel substitution, and policy options for reducing transportation-related greenhouse gas (GHG) emissions and petroleum use.

eBook ISBN: 978-1-62808-583-9
Retail Price: $250

Transition to Hydrogen Fuel Cell Vehicles

Editor: Selim Koca

Series: Energy Policies, Politics and Prices

Book Description: The authors of this book analyze the hydrogen infrastructure analysis and deployment scenarios, policy options for supporting hydrogen energy infrastructure and vehicle developments during the transition to they hydrogen fuel cell vehicles, and the costs of implementing selected policy options to encourage the transition to hydrogen.

Hardcover ISBN: 978-1-60741-806-1
Retail Price: $120